D1187924

Natural Computing Series

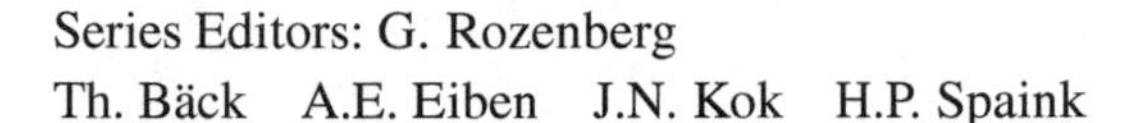

More information about this series at http://www.springer.com/series/4190

A.E. Eiben • J.E. Smith

Introduction to Evolutionary Computing

Second Edition

A.E. Eiben
Department of Computer Science
VU University Amsterdam
Amsterdam, The Netherlands

J.E. Smith
Department of Computer Science
and Creative Technologies
The University of the West of England
Bristol, UK

Series Editors
G. Rozenberg (Managing Editor)

Th. Bäck, J.N. Kok, H.P. Spaink
Leiden Center for Natural Computing
Leiden University
Leiden, The Netherlands

A.E. Eiben
VU University Amsterdam
The Netherlands

ISSN 1619-7127
Natural Computing Series
ISBN 978-3-662-44873-1 ISBN 978-3-662-44874-8 (eBook)
DOI 10.1007/978-3-662-44874-8

Library of Congress Control Number: 2015944812

Springer Heidelberg New York Dordrecht London

Printed on acid-free paper

Springer-Verlag GmbH Berlin Heidelberg is part of Springer Science+Business Media (www.springer.com)

Preface

This is the second edition of our 2003 book. It is primarily a book for lecturers and graduate and undergraduate students. To this group the book offers a thorough introduction to evolutionary computing (EC), descriptions of popular evolutionary algorithm (EA) variants, discussions of methodological issues and particular EC techniques. We end by presenting an outlook to evolutionary robotics and the future of EC, as it stands poised to make a major transition from evolution within computers to the evolution of things [147].

This book is also meant for those who wish to apply EC to a particular problem or within a given application area. To this group the book is valuable because it presents EC as something to be *used*, rather than just studied, and it contains an explicit treatment of guidelines for good experimentation. Finally, this book contains information on the state of the art in a wide range of subjects that are interesting to fellow researchers, as quick reference on subjects outside of their own specialist field of EC.

This book has a supporting website at

`www.evolutionarycomputation.org`

which offers additional information. In particular, the educational role of the book is emphasised:

1. There are exercises and a list of recommended further reading for each chapter.
2. The outline of a full academic course based on this book is given.
3. There are slides for each chapter in PDF and PowerPoint format. These slides can be freely downloaded, altered, and used to teach the material covered in the book.
4. Furthermore, the website offers answers to the exercises, downloadables for easy experimentation, a discussion forum, and errata.

When updating the book we altered its main logic. In the first edition, popular evolutionary algorithm variants, such as genetic algorithms or evolution strategies, had a prominent role. They were treated in separate chapters and specific representations and evolutionary operators were presented within the framework of one of these algorithm variants. In the second edition we are emphasising the generic scheme of EAs as an approach to problem-solving. This is reflected by the following major changes:

- We added a chapter on problems. Since the whole book is about problem solvers, we felt it was good to start with a chapter on problems.
- The treatment of EAs is organised according to the main algorithm components, such as representation, variation and selection operators.
- The most popular EA variants are presented as special cases of the generic EA scheme. Although the treatment of each variant is now shorter, the list of variants is longer, now including differential evolution, particle swarm optimisation, and estimation of distribution algorithms.

We also extended the treatment of the how-to parts of the book. We added a new chapter on parameter tuning and grouped this with the chapters on parameter control and the how-to-work-with content into a methodological part. Furthermore, we dropped the Exercises and Recommended Reading sections at the end of each chapter as they were too static. Instead, we offer these on the website for the book.

The overall structure of the new edition is three-tier: Part I presents the basics, Part II is concerned with methodological issues, and Part III discusses advanced topics. These parts are followed by the References, and although that now contains nearly five hundred entries, we inevitably missed some. We apologise, it is nothing personal. Just send us an email if we forgot a really important one.

Writing this book would not have been possible without the support of many. In the first place, we wish to express our gratitude to Daphne and Cally for their patience, understanding, and tolerance. Without their support this book could not have been written. Furthermore, we acknowledge the help of our colleagues and the students worldwide who pointed out errors in and gave us feedback about the earlier version of the book. We are especially grateful to Bogdan Filipič for his comments on the almost-final draft of this book.

We wish everybody a pleasant and fruitful time reading and using this book.

Amsterdam, Bristol, April 2015 *Gusz Eiben and Jim Smith*

Contents

Part I The Basics

Part I

The Basics

1

Problems to Be Solved

In this chapter we discuss problems to be solved, as encountered frequently by engineers, computer scientists, etc. We argue that problems and problem solvers can, and should, be distinguished, and observe that the field of evolutionary computing is primarily concerned with problem solvers. However, to characterise any problem solver it is useful to identify the kind of problems to which it can be applied. Therefore we start this book by discussing various classes of problems, and, in fact, even different ways of classifying problems.

In the following informal discussion, we introduce the concepts and the terminology needed for our purposes by examples, only using a formal treatment when it is necessary for a good understanding of the details. To avoid controversy, we are not concerned with social or political problems. The problems we have in mind are the typical ones with which artificial intelligence is associated: more akin to puzzles (e.g., the famous zebra puzzle), numerical problems (e.g., what is the shortest route from a northern city to a southern city), or pattern discovery (e.g., what will a new customer buy in our online book store, given their gender, age, address, etc).

1.1 Optimisation, Modelling, and Simulation Problems

The classification of problems used in this section is based on a black box model of computer systems. Informally, we can think of any computer-based system as follows. The system initially sits, awaiting some input from either a person, a sensor, or another computer. When input is provided, the system processes that input through some computational model, whose details are not specified in general (hence the name black box). The purpose of this model is to represent some aspects of the world relevant to the particular application. For instance, the model could be a formula that calculates the total route length from a list of consecutive locations, a statistical tool estimating the likelihood of rain given some meteorological input data, a mapping from real-time data regarding a car's speed to the level of acceleration necessary to approach some prespecified target speed, or a complex series of rules that transform a series of keystrokes into an on screen version of the page you are reading now. After processing the input the system provides some outputs – which might be messages on screen, values written to a file, or commands sent to an actuator such as an engine. Depending on the application, there might be one or more inputs of different types, and the computational model might be simple, or very complex. Importantly, knowing the model means that we can compute the output for any input. To provide some concrete examples:

- When designing aircraft wings, the inputs might represent a description of a proposed wing shape. The model might contain equations of complex fluid dynamics to estimate the drag and lift coefficients of any wing shape. These estimates form the output of the system.
- A voice control system for smart homes takes as input the electrical signal produced when a user speaks into a microphone. Suitable outputs might be commands to be sent to the heating system, the TV set, or the lights. Thus in this case the model consists of a mapping from certain patterns in electrical waveforms coming from an audio input onto the outputs that would normally be created by key-presses on a keyboard.
- For a portable music player, the inputs might be a series of gestures and button presses – perhaps choosing a playlist that the user has created. Here the response of the model might involve selecting a series of mp3 files from a database and processing them in some way to provide the desired output for that sequence of gestures. In this case the output would be a fluctuating electrical signal fed to a pair of earphones that in turn produce the sound of the chosen songs.

In essence, the black box view of systems distinguishes three components, the input, the model, and the output. In the following we will describe three problem types, depending on which of these three is unknown.

1.1.1 Optimisation

In an **optimisation** problem the model is known, together with the desired output (or a description of the desired output), and the task is to find the input(s) leading to this output (Fig. 1.1).

For an example, let us consider the travelling salesman problem. This apparently rather abstract problem is popular in computer science, as there are many practical applications which can be reduced to this, such as organising delivery routes, plant layout, production schedules, and timetabling. In the abstract version we are given a set of cities and have to find the shortest tour which visits each city exactly once. For a given instance of this problem, we have a formula (the model) that for each given sequence of cities (the inputs) will compute the length of the tour (the output). The problem is to find an input with a desired output, that is, a sequence of cities with optimal (minimal) length. Note that in this example the desired output is defined implicitly.

That is, rather specifying the exact length, it is required that the tour should be shorter than all others, and we are looking for inputs realising this.

Another example is that of the eight-queens problem. Here we are given a chess board and eight queens that need to be placed on the board in such a way that no two queens can check each other, i.e., they must not share the same row, column, or diagonal. This problem can be captured by a computational system where an input is a certain configuration of all eight queens, the model calculates whether the queens in a given configuration check each other or not, and the output is the number of queens not being checked. As opposed to the travelling salesman problem, here the desired output is specified explicitly: the number of queens not being checked must be eight. An alternative system capturing this problem could have the same set of inputs, the same model, but the output can be a simple binary value, representing "OK" or "not OK", referring to the configuration as a whole. In this case we are looking for an input that generates "OK" as output. Intuitively, this problem may not feel like real optimisation, because there is no graded measure of goodness. In Sect. 1.3 we will discuss this issue in more detail.

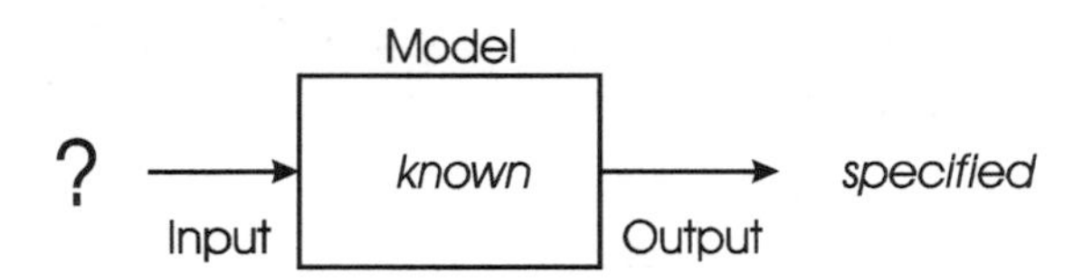

Fig. 1.1. Optimisation problems. These occur frequently in engineering and design. The label on the Output reads "specified", instead of "known", because the specific value of the optimum may not be known, only defined implicitly (e.g., the lowest of all possibilities).

1.1.2 Modelling

In a **modelling** or **system identification** problem, corresponding sets of inputs and outputs are known, and a model of the system is sought that delivers the correct output for each known input (Fig. 1.2). In terms of human learning this corresponds to finding a model of the world that matches our previous experience, and can hopefully generalise to as-yet unseen examples.

Let us take the stock exchange as an example, where some economic and societal indices (e.g., the unemployment rate, gold price, euro–dollar exchange rate, etc.) form the input, and the Dow Jones index is seen as output. The task is now to find a formula that links the known inputs to the known outputs, thereby representing a model of this economic system. If one can find a correct model for the known data (from the past), and if we have good reasons to believe that the relationships captured in this model remain true, then we have a prediction tool for the value of the Dow Jones index given new data.

As another example, let us take the task of identifying traffic signs in images – perhaps from video feeds in a smart car. In this case the system is composed of two elements. In a preprocessing stage, image processing routines take the electrical signals produced by the camera, divide these into regions of interest that might be traffic signs, and for each one they produce a set of numerical descriptors of the size, shape, brightness, contrast, etc. These values represent the image in a digital form and we consider the preprocessing component to be given for now. Then in the main system each input is a vector of numbers describing a possible sign, and the corresponding output is a label from a predefined set, e.g., "stop", "give-way", "50", etc. (the traffic sign). The model is then an algorithm which takes images as input and produces labels of traffic signs as output. The task here is to produce a model that responds with the appropriate traffic sign labels in every situation. In practice, the set of all possible situations would be represented by a large collection of images that are all labelled appropriately. Then the modelling problem is reduced to finding a model that gives a correct output for each image in the collection.

Also the voice control system for smart homes described in the beginning of this section includes a modelling problem. The set of all phrases pronounced by the user (inputs) must be correctly mapped onto the set of all control commands in the repertoire of the smart home.

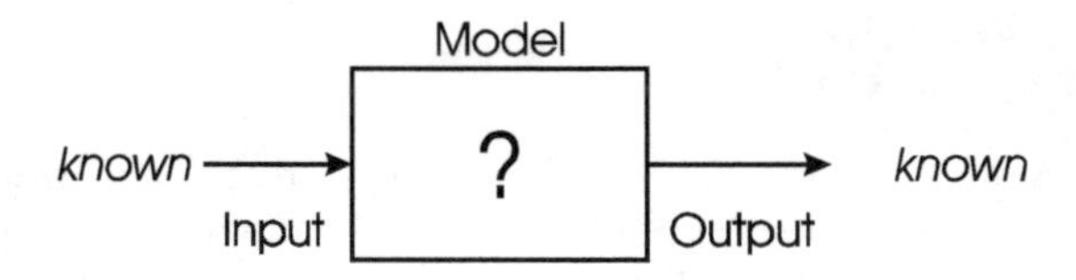

Fig. 1.2. Modelling or system identification problems. These occur frequently in data mining and machine learning

It is important to note that modelling problems can be transformed into optimisation problems. The general trick is to designate the error rate of a model as the quantity to be minimised or its hit rate to be maximised. As an example, let us take the traffic sign identification problem. This can be formulated as a modelling problem: that of finding the correct model m that maps each one of a collection of images onto the appropriate label(s) identifying the traffic signs in that image. The model m that solves the problem is unknown in advance, hence the question mark in Figure 1.2. In order to find a solution we need to start by choosing a technology. For instance, we may wish to have it as a decision tree, an artificial neural network, a piece of Java code, or a MATLAB expression. This choice allows us to specify the required form or syntax of m. Having done that, we can define the set of all possible solutions M for our chosen technology, being all correct expressions in the given syntax, e.g., all decision trees with the appropriate variables or all possible artificial neural networks with a given topology. Now we can define a related optimisation problem. The set of inputs is M and the output for a given $m \in M$ is an integer saying how many images were correctly labelled by m. It is clear that a solution of this optimisation problem with the maximum number of correctly labelled images is a solution to the original modelling problem.

1.1.3 Simulation

In a **simulation** problem we know the system model and some inputs, and need to compute the outputs corresponding to these inputs (Fig. 1.3). As an example, think of an electronic circuit, say, a filter cutting out low frequencies in a signal. Our model is a complex system of formulas (equations and inequalities) describing the working of the circuit. For any given input signal this model can compute the output signal. Using this model (for instance, to compare two circuit designs) is much cheaper than building the circuit and measuring its properties in the physical world. Another example is that of a weather forecast system. In this case, the inputs are the meteorological data regarding, temperature, wind, humidity, rainfall, etc., and the outputs are actually the same: temperature, wind, humidity, rainfall, etc., but at a different time. The model here is a temporal one to predict meteorological data.

Simulation problems occur in many contexts, and using simulators offers various advantages in different applications. For instance, simulation can be more economical than studying the real-world effects, e.g., for the electronic circuit designers. The real-world alternative may not be feasible at all, for instance, performing what-if analyses of various tax systems *in vivo* is practically impossible. And simulation can be the tool that allows us to look into the future, as in weather forecast systems.

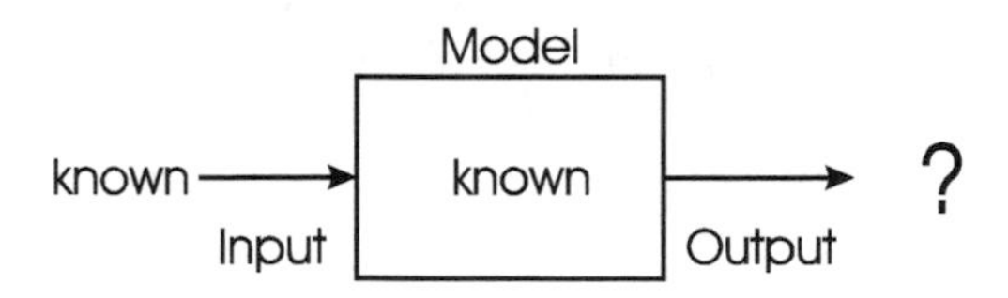

Fig. 1.3. Simulation problems. These occur frequently in design and in socio-economical contexts

1.2 Search Problems

A deeply rooted assumption behind the black box view of systems is that a computational model is directional: it computes from the inputs towards the outputs and it cannot be simply inverted. This implies that solving a simulation problem is different from solving an optimisation or a modelling problem. To solve a simulation problem, we only need to apply the model to some inputs and simply wait for the outcome.[1] However, solving an optimisation or a modelling problem requires the identification of a particular object in a space of possibilities. This space can be, and usually is, enormous. This leads us to the notion that the process of problem solving can be viewed as a search through a potentially huge set of possibilities to find the desired solution. Consequently, the problems that are to be solved this way can be seen as search problems. In terms of the classification of problems discussed in Section 1.1, optimisation and modelling problems can be naturally perceived as search problems, while this does not hold for simulation problems.

This view naturally leads to the concept of a **search space**, being the collection of all objects of interest including the solution we are seeking. Depending on the task at hand, the search space consists of all possible inputs to a model (optimisation problems), or all possible computational models that describe the phenomenon we study (modelling problems). Such search spaces can indeed be very large; for instance, the number of different tours through n cities is $(n-1)!$, and the number of decision trees with real-valued parameters is infinite. The specification of the search space is the first step in defining a search problem. The second step is the definition of a solution. For optimisation problems such a definition can be explicit, e.g., a board configuration where the number of checked queens is zero, or implicit, e.g., a tour that is the shortest of all tours. For modelling problems, a solution is defined by the property that it produces the correct output for every input. In practice, however, this is often relaxed, only requiring that the number of inputs for which the output is correct be maximal. Note that this approach transforms the modelling problem into an optimisation one, as illustrated in Section 1.1.2.

This notion of problem solving as search gives us an immediate benefit: we can draw a distinction between (search) problems – which define search spaces – and problem solvers – which are methods that tell us how to move through search spaces.

1.3 Optimisation Versus Constraint Satisfaction

The classification scheme discussed in this section is based on distinguishing between objective functions to be optimised and constraints to be satisfied. In general, we can consider an **objective function** to be some way of assigning a value to a possible solution that reflects its quality on a scale, whereas a **constraint** represents a binary evaluation telling us whether a given requirement holds or not. In the previous sections several objective functions were mentioned, including:

[1] The main challenge here is very often to build the simulator, which, in fact, amounts to solving a modelling problem.

(1) the number of unchecked queens on a chess board (to be maximised);
(2) the length of a tour visiting each city in a given set exactly once (to be minimised);
(3) the number of images in a collection that are labelled correctly by a given model m (to be maximised).

These examples illustrate that solutions to a problem can be identified in terms of optimality with respect to some objective function. Additionally, solutions can be subject to constraints phrased as criteria that must be satisfied. For instance:

(4) Find a configuration of eight queens on a chess board such that no two queens check each other.
(5) Find a tour with minimal length for a travelling salesman such that city X is visited after city Y.

There are a number of observations to be made about these examples. Example 2 refers to a problem whose solution is defined purely in terms of optimisation. On the other hand, example 4 illustrates the case where a solution is defined solely in terms of a constraint: a given configuration is either good or not. Note that this overall constraint regarding a whole configuration is actually composed from more elementary constraints concerning pairs of queens. A complete configuration is OK if all pairs of queens are OK. Example 5 is a mixture of these two basic types since it has an objective function (tour length) and a constraint (visit X after Y). Based on these observations we can set up another system for classifying problems, depending on the presence or absence of an objective function and constraints in the problem definition. The resulting four categories are shown in Table 1.1.

	Objective function	
Constraints	Yes	No
Yes	Constrained optimisation problem	Constraint satisfaction problem
No	Free optimisation problem	No problem

Table 1.1. Problem types distinguished by the presence or absence of an objective function and constraints

In these terms, the travelling salesman problem (item 2 above) is a **free optimisation problem (FOP)**, the eight-queens problem (item 4 above) is a **constraint satisfaction problem (CSP)**, and the problem shown in item 5 is a **constrained optimisation problem (COP)**. Comparing items 1 and 4 we can see that constraint satisfaction problems can be transformed into optimisation problems. The basic trick is the same as in transforming modelling problems into optimisation problems: rather than requiring perfection, we just count the number of satisfied constraints (e.g., non-checking pairs of queens) and introduce this as an objective function to be maximised. Obviously, an object (e.g., a board configuration) is a solution of the original constraint satisfaction problem if and only if it is a solution of this associated optimisation problem.

To underpin further interesting insights about problems, let us have a closer look at the eight-queens problem. Its original formulation is in natural language:

> Place eight queens on a chess board in such a way that no two queens check each other.

This problem definition is informal in the sense that it lacks any reference to the formal constructs we have introduced here, such as inputs/outputs, a search space, etc. In order to develop an algorithm for this problem, it needs to be formalised. As it happens, it can be formalised in different ways, and these lead to different types of formal problems describing it. The easiest way to illustrate a number of options is to take the search perspective.

FOP If we define search space S to be the set of all board configurations with eight queens, we can capture the original problem as a free optimisation problem with an objective function f that reports the number of free queens for a given configuration, and define a solution as a configuration $s \in S$ with $f(s) = 8$.

CSP Alternatively, we can formalise it as a constraint satisfaction problem with the same search space S and define a constraint ϕ such that $\phi(s) = true$ if and only if no two queens check each other for the configuration s.

COP Yet another formalisation is obtained if we take a different search space. This can be motivated by the observation that in any solution of the eight-queens problem the number of queens in each column must be exactly one. Obviously, the same holds for rows. So we could distinguish vertical constraints (for columns), horizontal constraints (for rows), and diagonal constraints, and decide to restrict ourselves to board configurations that satisfy the vertical and horizontal constraints already. This is a workable approach, since it is rather easy to find configurations with one queen in each column and in each row. These configurations are a subset of the original search space – let us call this S'. Formally, we can then define a constrained optimisation problem over S with a modified constraint ψ' such that $\psi'(s) = true$ if and only if all vertical and horizontal constraints are satisfied in s (i.e. $\phi'(s) = true$ if and only if s is in S') and a new function g that reports the number of pairs of queens in s that violate the diagonal constraints. It is easy to see that a board configuration is a solution of the eight-queens problem if and only if it is a solution of this constrained optimisation problem with $g(s) = 0$ and $\phi'(s) = true$.

These examples illustrate that the nature of a problem is less obvious than it may seem. In fact, it all depends on how we choose to formalise it. Which formalisation is to be preferred is a subject for discussion. It can be argued that some formalisations are more natural, or fit the problem better, than others. For instance, one may prefer to see the eight-queens problem as a constraint satisfaction problem by nature and consider all other formalisations as secondary transformations. Likewise, one can consider the traffic sign recognition problem as a modelling problem in the first place and transform it to an optimisation problem for practical purposes. Algorithmic considerations can also be a major influence here. If one has an algorithm that can solve free optimisation problems well, but cannot cope with constraints, then it is very sensible to formalise problems as free optimisation.

1.4 The Famous NP Problems

Up to this point we have discussed a number of different ways of categorising problems, and have deliberately stayed away from discussions about problem-solvers. Consequently, it is possible to classify a problem according to one of those schemes by only looking at the problem. In this section we discuss a classification scheme where this is not possible because the problem categories are defined through the properties of problem-solving algorithms. The motivation behind this approach is the intention to talk about problems in terms of their difficulty, for instance, being hard or easy to solve. Roughly speaking, the basic idea is to call a problem easy if there exists a fast solver for it, and hard otherwise. This notion of problem hardness leads to the study of computational complexity.

Before we proceed we need to make a further distinction among optimisation problems, depending on the type of objects in the corresponding search space. If the search space S is defined by continuous variables (i.e., real numbers), then we have a **numerical optimisation problem**. If S is defined by discrete variables (e.g., Booleans or integers), then we have a **combinatorial optimisation problem**. The various notions of problem hardness discussed further on are defined for combinatorial optimisation problems. Notice that discrete search spaces are always finite or, in the worst case, countably infinite.

We do not attempt to provide a complete overview of computational complexity as this is well covered in many books, such as [180, 330, 331, 318]. Rather, we provide a brief outline of some important concepts, their implications for problem-solving, and also of some very common misconceptions. Furthermore, we do not treat the subject with mathematical rigour as it would not be appropriate for this book. Thus, we do not give precise definitions of essential concepts, like algorithm, problem size, or run-time, but use such terms in an intuitive manner, explaining their meaning by examples if necessary.

The first key notion in computational complexity is that of **problem size**, which is grounded in the dimensionality of the problem at hand (i.e., the number of variables) and the number of different values for the problem variables. For the examples discussed before, the number of cities to visit, or the number of queens to place on the board could be sensible measures to indicate problem size. The second notion concerns algorithms, rather than problems. The **running-time** of an algorithm is the number of elementary steps, or operations, it takes to terminate. The general, although not always correct, intuition behind computational complexity is that larger problems need more time to solve. The best-known definitions of problem hardness relate the size of a problem to the (worst-case) running-time of an algorithm to solve it. This relationship is expressed by a formula that specifies an upper-bound for the worst-case running-time as a function of the problem size. To put it simply, this formula can be polynomial (considered to indicate relatively short running-times) or superpolynomial, e.g., exponential (indicating long running-times). The final notion is that of **problem reduction**, which is the idea that we can transform one problem into another via a suitable mapping. Note that the transformation might not be reversible. Although this idea of transforming or reducing problems is slightly complex, it is not entirely unfamiliar since we saw in the previous section that a given problem in the real world can often by formalised in different, but equivalent ways. The frequently used notions regarding problem hardness can now be phrased as follows.

A problem is said to belong to the **class P** if there exists an algorithm that can solve it in polynomial time. That is, if there exists an algorithm for it whose worst-case running-time for problem size n is less than $F(n)$ for some polynomial formula F. In common parlance, the set P contains the problems that can be easily solved, e.g., the Minimum Spanning Tree problem.

A problem is said to belong to the **class NP** if it can be solved by some algorithm (with no claims about its run-time) and any solution can be verified within polynomial time by some other algorithm.[2] Note that it follows that P is a subset of NP, since a polynomial solver can also be used to verify solutions in polynomial time. An example of an NP-problem is the subset-sum problem: given a set of integers, is there some set of one or more elements of that set which sum to zero? Clearly, giving a negative answer to this problem for a given set of numbers would require examining all possible subsets. Unfortunately, the number of the possible subsets is more than polynomial in the size of the set. However verifying that a solution is valid merely involves summing the contents of the subset discovered.

A problem is said to belong to the **class NP-complete** if it belongs to the class NP and any other problem in NP can be reduced to this problem by an algorithm which runs in polynomial time. In practice these represent difficult problems which crop up all the time. Several large lists of well-known examples of NP-complete problems can readily be found on the internet –

[2] For the sake of correctness, here we commit the most blatant oversimplification. We 'define' NP without any reference to non-deterministic Turing Machines, or restricting the notion to decision problems.

we will not attempt to summarise other than to say that the vast majority of interesting problems in computer science turn out to be NP-complete.

Finally a problem is said to belong to the **class NP-hard** if it is at least as hard as any problem in NP-complete (so all problems in NP-complete can be reduced to one in NP-hard), but where the solutions cannot necessarily be verified within polynomial time. One such example is the Halting Problem.

The existence of problems where a solution cannot be verified in polynomial time proves that the class P is not the same as the class NP-hard. What is unknown is whether the two classes P and NP are in fact the same. If this were to be the case then the implications would be enormous for computer science and mathematics as it would be known that fast algorithms must exist for problems which were previously thought to be difficult. Thus whether $P = NP$ is one of the grand challenges of complexity theory, and there is a million-dollar reward offered for any proof that $P = NP$ or $P \neq NP$. Notice, that while the latter is the subject of much complex mathematics, the former could simply be proved by the creation of a fast algorithm for any of the NP-complete problems, for instance, an algorithm for the travelling salesman problem whose worst-case running-time scaled polynomially with the number of cities. Figure 1.4 shows the classes of problem hardness depending on the equality of P and NP. If $P = NP$ then the sets $P = NP = NP$-complete but they are still a subset of NP-hard.

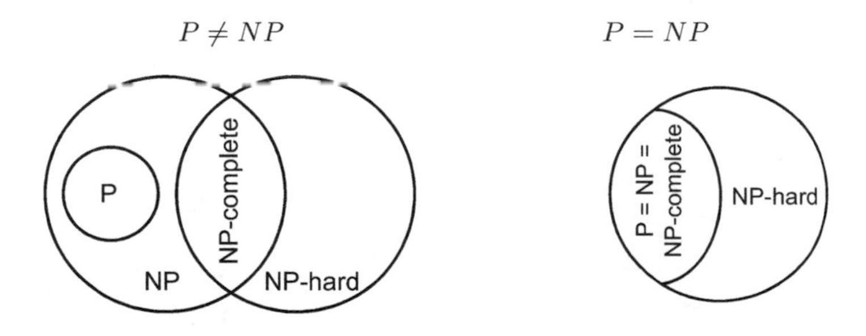

Fig. 1.4. Classes of problem hardness depending on the equality of P and NP

While this sounds rather theoretical, it has some very important implications for problem-solving. If a problem is NP-complete, then although we might be able to solve particular instances within polynomial time, we cannot say that we will be able to do so for all possible instances. Thus if we wish to apply problem-solving methods to those problems we must currently either accept that we can probably only solve very small (or otherwise easy) instances, or give up the idea of providing exact solutions and rely on approximation or metaheuristics to create good enough solutions. This is in contrast to problems which are known to be in P. Although the number of possible solutions for these problems may scale exponentially, algorithms exist which find solutions and whose running-times scale polynomially with the size of the instance.

To summarise this section, there are a huge number of practical problems which, on examination, turn out to be a variant of an abstract problem that is known to be in the class NP-complete. Although some instances of such a problem might be easy, most computer scientists believe that no polynomial-time algorithm exists for such problems, and certainly one has not yet been discovered. Therefore, if we wish to be able to create acceptable solutions for any instance of such a problem, we must turn to the use of approximation and metaheuristics and abandon the idea of definitely finding a solution which is provably the best for the instance.

For exercises and recommended reading for this chapter, please visit `www.evolutionarycomputation.org`.

2

Evolutionary Computing: The Origins

This chapter provides the reader with the basics for studying evolutionary computing (EC) through this book. We begin with a brief history of the field of evolutionary computing, followed by an introduction to some of the biological processes that have served as inspiration and that have provided a rich source of ideas and metaphors to researchers. We then discuss motivations for working with, and studying, evolutionary computing methods. We end with examples of applications where EC was successfully applied.

2.1 The Main Evolutionary Computing Metaphor

Evolutionary computing is a research area within computer science. As the name suggests, it is a special flavour of computing, which draws inspiration from the process of natural evolution. It is not surprising that some computer scientists have chosen natural evolution as a source of inspiration: the power of evolution in nature is evident in the diverse species that make up our world, each tailored to survive well in its own niche. The fundamental metaphor of evolutionary computing relates this powerful natural evolution to a particular style of problem solving – that of trial-and-error.

Descriptions of relevant fragments of evolutionary theory and genetics are given later on. For the time being let us consider natural evolution simply as follows. A given environment is filled with a population of individuals that strive for survival and reproduction. The fitness of these individuals is determined by the environment, and relates to how well they succeed in achieving their goals. In other words, it represents their chances of survival and of multiplying. Meanwhile, in the context of a stochastic trial-and-error (also known as generate-and-test) style problem solving process, we have a collection of candidate solutions. Their quality (that is, how well they solve the problem) determines the chance that they will be kept and used as seeds for constructing further candidate solutions. The analogies between these two scenarios are shown in Table 2.1.

Evolution		Problem solving
Environment	$\longleftrightarrow$	Problem
Individual	$\longleftrightarrow$	Candidate solution
Fitness	$\longleftrightarrow$	Quality

Table 2.1. The basic evolutionary computing metaphor linking natural evolution to problem solving

2.2 Brief History

Surprisingly enough, this idea of applying Darwinian principles to automated problem solving dates back to the 1940s, long before the breakthrough of computers [167]. As early as 1948, Turing proposed "genetical or evolutionary search", and by 1962 Bremermann had actually executed computer experiments on "optimization through evolution and recombination". During the 1960s three different implementations of the basic idea were developed in different places. In the USA, Fogel, Owens, and Walsh introduced **evolutionary programming** [173, 174], while Holland called his method a **genetic algorithm** [102, 218, 220]. Meanwhile, in Germany, Rechenberg and Schwefel invented **evolution strategies** [352, 373]. For about 15 years these areas developed separately; but since the early 1990s they have been viewed as different representatives ('dialects') of one technology that has come to be known as **evolutionary computing** (EC) [22, 27, 28, 137, 295, 146, 104, 12]. In the early 1990s a fourth stream following the general ideas emerged, **genetic programming**, championed by Koza [37, 252, 253]. The contemporary terminology denotes the whole field by evolutionary computing, the algorithms involved are termed **evolutionary algorithms**, and it considers evolutionary programming, evolution strategies, genetic algorithms, and genetic programming as subareas belonging to the corresponding algorithm variants.

The development of scientific forums devoted to EC gives an indication of the field's past and present, and is sketched in Fig. 2.1. The first international conference specialising in the subject was the *International Conference on Genetic Algorithms* (ICGA), first held in 1985 and repeated every second year until 1997. In 1999 it merged with the *Annual Conference on Genetic Programming* to become the annual *Genetic and Evolutionary Computation Conference* (GECCO). At the same time, in 1999, the *Annual Conference on Evolutionary Programming*, held since 1992, merged with the *IEEE Conference on Evolutionary Computation*, held since 1994, to form the *Congress on Evolutionary Computation* (CEC), which has been held annually ever since.

The first European event (explicitly set up to embrace all streams) was *Parallel Problem Solving from Nature* (PPSN) in 1990, which has became a biennial conference. The first scientific journal devoted to this field, *Evolutionary Computation*, was launched in 1993. In 1997 the European Commission decided to fund a European research network in EC, called EvoNet, whose

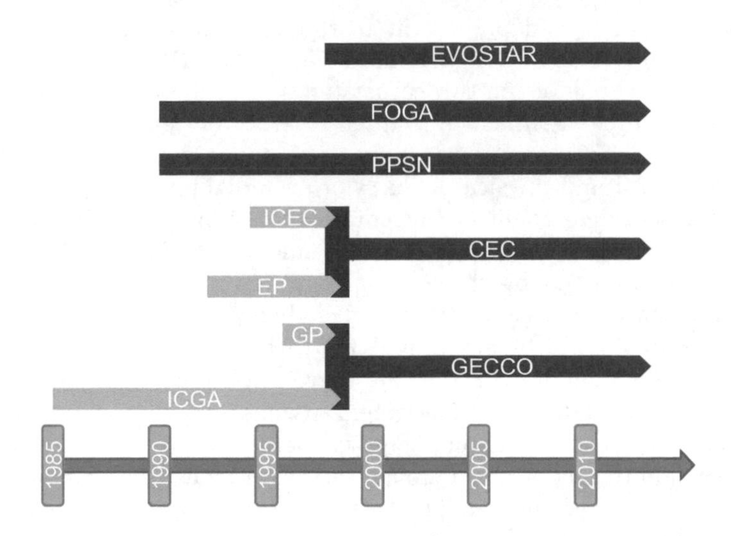

Fig. 2.1. Brief sketch of the EC conference history

funds were guaranteed until 2003. At the time of writing (2014), there were three major EC conferences (CEC, GECCO, and PPSN) and many smaller ones, including one dedicated exclusively to theoretical analysis and development, *Foundations of Genetic Algorithms* (FOGA), held biennially since 1990, and a European event seeded by EvoNet, the annual EVOSTAR conference. There are now various scientific EC journals (*Evolutionary Computation*, *IEEE Transactions on Evolutionary Computation*, *Genetic Programming and Evolvable Machines*, *Evolutionary Intelligence*, *Swarm and Evolutionary Computing*) and many with a closely related profile, e.g., on natural computing, soft computing, or computational intelligence. We estimate the number of EC publications in 2014 at somewhere over 2000 – many of them in journals and conference proceedings of specific application areas.

2.3 The Inspiration from Biology

2.3.1 Darwinian Evolution

Darwin's theory of evolution [92] offers an explanation of the origins of biological diversity and its underlying mechanisms. In what is sometimes called the macroscopic view of evolution, natural selection plays a central role. Given an environment that can host only a limited number of individuals, and the basic instinct of individuals to reproduce, selection becomes inevitable if the population size is not to grow exponentially. Natural selection favours those individuals that compete for the given resources most effectively, in other words, those that are adapted or fit to the environmental conditions best. This phenomenon is also known as **survival of the fittest.**[1] Competition-based selection is one of the two cornerstones of evolutionary progress. The other primary force identified by Darwin results from phenotypic variations among members of the population. Phenotypic traits (see also Sect. 2.3.2) are those behavioural and physical features of an individual that directly affect its response to the environment (including other individuals), thus determining its fitness. Each individual represents a unique combination of phenotypic traits that is evaluated by the environment. If this combination evaluates favourably, then the individual has a higher chance of creating offspring; otherwise the individual is discarded by dying without offspring. Importantly, if they are heritable (and not all traits are), favourable phenotypic traits may be propagated via the individual's offspring. Darwin's insight was that small, random variations – mutations – in phenotypic traits occur during reproduction from generation to generation. Through these variations, new combinations of traits occur and get evaluated. The best ones survive and reproduce, and so evolution progresses. To summarise this basic model, a population consists of a number of individuals. These individuals are the units of selection, that is to say that their reproductive success depends on how well they are adapted to their environment relative to the rest of the population. As the more successful individuals reproduce, occasional mutations give rise to new individuals to be tested. Thus, as time passes, there is a change in the constitution of the population, i.e., the population is the unit of evolution.

This process is well captured by the intuitive metaphor of an **adaptive landscape** or adaptive surface [468]. On this landscape the height dimension belongs to fitness: high altitude stands for high fitness. The other two (or more, in the general case) dimensions correspond to biological traits as shown in Fig. 2.2. The xy-plane holds all possible trait combinations, and the z-values show their fitnesses. Hence, each peak represents a range of successful trait combinations, while troughs belong to less fit combinations. A given population can be plotted as a set of points on this landscape, where each dot is one individual realising a possible trait combination. Evolution is then the process of gradual advances of the population to high-altitude areas, powered by variation and natural selection. Our familiarity with the physical landscape on which we exist naturally leads us to the concept of **multimodal problems**. These are problems in which there are a number of points that are better than all their neighbouring solutions. We call each of these points a **local optimum** and denote the highest of these as the **global optimum**. A problem in which there is only one local optimum is known as **unimodal**.

The link with an optimisation process is as straightforward as it is misleading, because evolution is not a unidirectional uphill process [103]. Because the

[1] This term is actually rather misleading. It is often, and incorrectly, taken to mean that the best fit individual *always* survives. Since nature, and EC by design, contains a lot of randomness, this does not always happen.

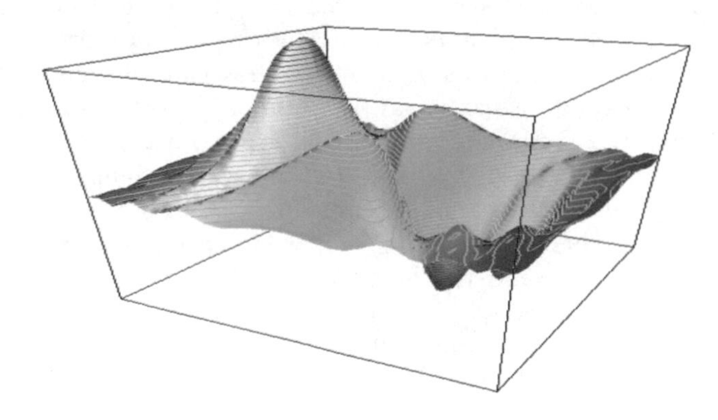

Fig. 2.2. Illustration of Wright's adaptive landscape with two traits

population has a finite size, and random choices are made in the selection and variation operators, it is common to observe the phenomenon of **genetic drift**, whereby highly fit individuals may be lost from the population, or the population may suffer from a loss of variety concerning some traits. This can have the effect that populations 'melt down' the hill, and enter low-fitness valleys. The combined global effects of drift and selection enable populations to move uphill as well as downhill, and of course there is no guarantee that the population will climb back up the same hill. Escaping from locally optimal regions is hereby possible, and according to Wright's shifting balance theory the maximum of a fixed landscape can be reached.

2.3.2 Genetics

The microscopic view of natural evolution is offered by molecular genetics. It sheds light on the processes below the level of visible phenotypic features, in particular relating to heredity. The fundamental observation from genetics is that each individual is a dual entity: its phenotypic properties (outside) are represented at a genotypic level (inside). In other words, an individual's **genotype** encodes its **phenotype**. **Genes** are the functional units of inheritance encoding phenotypic characteristics. For instance, visible properties like the fur colour or tail length could be determined by genes. Here it is important to distinguish genes and **alleles**. An allele is one of the possible values that a gene can have – so its relationship to a gene is just like that of a specific value to a variable in mathematics. To illustrate this by an oversimplified example, bears could have a gene that determines fur colour, and for a polar bear we would expect to see the allele that specifies the colour white. In natural systems the genetic encoding is not one-to-one: one gene might affect more phenotypic traits (**pleitropy**), and in turn one phenotypic trait can be determined by more than one gene (**polygeny**). Phenotypic variations are always caused by genotypic variations, which in turn are the consequences of mutations of genes, or recombination of genes by sexual reproduction.

Another way to think of this is that the genotype contains all the information necessary to build the particular phenotype. The term **genome** stands for the complete genetic information of a living being containing its total building plan. This genetic material, that is, all genes of an organism, is arranged in several chromosomes; there are 46 in humans. Higher life forms (many plants and animals) contain a double complement of chromosomes in most of their cells, and such cells – and the host organisms – are called **diploid**. Thus the chromosomes in human diploid cells are arranged into 23 pairs. **Gametes** (i.e., sperm and egg cells) contain only one single complement of chromosomes and are called **haploid**. The combination of paternal and maternal features in the offspring of diploid organisms is a consequence of fertilisation by a fusion of such gametes: the haploid sperm cell merges with the haploid egg cell and forms a diploid cell, the zygote. In the zygote, each chromosome pair is formed by a paternal and a maternal half. The new organism develops from this zygote by the process named **ontogenesis**, which does not change the genetic information of the cells. Consequently, all body cells of a diploid organism contain the same genetic information as the zygote it originates from.

In evolutionary computing, the combination of features from two individuals in offspring is often called crossover. It is important to note that this is not analogous to the working of diploid organisms, where **crossing-over** is not a process during mating and fertilisation, but rather happens during the formation of gametes, a process called meiosis. **Meiosis** is a special type of cell division that ensures that gametes contain only one copy of each chromosome. As said above, a diploid body cell contains chromosome pairs, where one half of the pair is identical to the paternal chromosome from the sperm cell, and the other half is identical to the maternal chromosome from the egg cell. During meiosis a chromosome pair first aligns physically, that is, the copies of the paternal and maternal chromosomes, which form the pair, move together and stick to each other at a special position (the centromere, not indicated, see Fig. 2.3, left). In the second step the chromosomes double so that four strands (called chromatids) are aligned (Fig. 2.3, middle). The actual crossing-over takes place between the two inner strands that break at a random point and exchange parts (Fig. 2.3, right). The result is four differ-

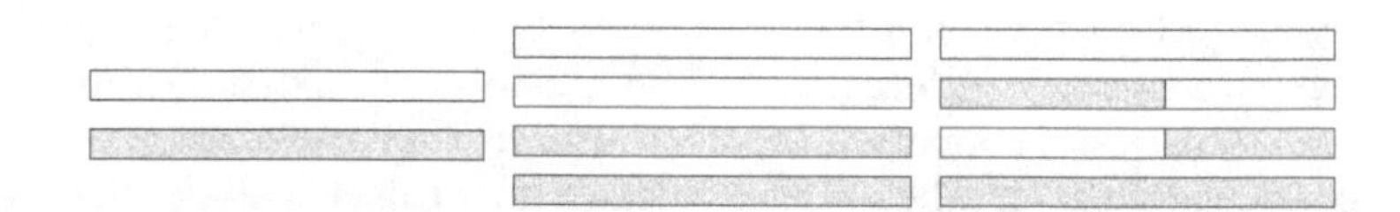

Fig. 2.3. Three steps in the (simplified) meiosis procedure regarding one chromosome

ent copies of the chromosome in question, of which two are identical to the original parental chromosomes, and two are new recombinations of paternal and maternal material. This provides enough genetic material to form four haploid gametes, which is done via a random arrangement of one copy of each chromosome. Thus in the newly created gametes the genome is composed of chromosomes that are either identical to one of the parent chromosomes, or recombinants. The resulting four haploid gametes are usually different from both original parent genomes, facilitating genotypic variation in offspring.

In the 19th century Mendel first investigated and understood heredity in diploid organisms. Modern genetics has added many details to his early picture, but we are still very far from understanding the whole genetic process. What we do know is that all life on Earth is based on DNA – the famous double helix of nucleotides encoding the whole organism be it a plant, animal, or *Homo sapiens*. Triplets of nucleotides form so-called codons, each of which codes for a specific amino acid. The genetic code (the translation table from the $4^3 = 64$ possible codons to the 20 amino acids from which proteins are created) is universal, that is, it is the same for all life on Earth. This fact is generally acknowledged as strong evidence that the whole biosphere has the same origin. Genes are larger structures on the DNA, containing many codons, carrying the code of proteins. The path from DNA to protein consists of two main steps: transcription, where information from the DNA is written to RNA, and translation, the step from RNA to protein. It is one of the principal dogmas of molecular genetics that this information flow is only one-way. Speaking in terms of genotypes and phenotypes, this means that phenotypic features cannot influence genotypic information. This refutes earlier theories (for instance, that of Lamarck), which asserted that features acquired during an individual's lifetime could be passed on to its offspring via inheritance. A consequence of this view is that changes in the genetic material of a population can only arise from random variations and natural selection and definitely not from individual learning. It is important to understand that all variations (mutation and recombination) happen at the genotypic level, while selection is based on actual performance in a given environment, that is, at the phenotypic level.

2.3.3 Putting It Together

The Darwinian theory of evolution and the insights from genetics can be put together to clarify the dynamics behind the emergence of life on Earth. For the purposes of this book a simplified picture is sufficient. The main points are then the following. Any living being is a dual entity with an invisible code (its genotype) and observable traits (its phenotype). Its success in surviving and reproducing is determined by its phenotypical properties, e.g., good ears, strong muscles, white fur, friendly social attitude, attractive scent, etc. In other words, the forces known as natural selection and sexual selection act on the phenotype level. Obviously, selection also affects the genotype level, albeit implicitly. The key here is reproduction. New individuals may have one single parent (**asexual reproduction**) or two parents (**sexual reproduction**). In either case, the genome of the new individual is not identical to that of the parent(s), because of small reproductive variations and because the combination of two parents will differ from both. In this way genotype variations are created, which in turn translate to phenotype variations[2] and thus are subject to selection. Hence, at a second level, genes are also subject to the game of survival and reproduction, and some evolutionary biologists would argue that viewing evolution from the perspective of genes is more productive – so that rather than thinking about populations of individuals, we should think about a 'gene pool' containing genes which compete and replicate over time, being evaluated as they reoccur in different individuals [100].

Elevating this process to an abstract level, we can perceive each newborn individual as a new sample in the space of all possible living things. This new sample is produced by forces of variation, i.e., asexual or sexual reproduction, and it is evaluated by the forces of selection. It needs to pass two hurdles: first proving viable to live on its own, then proving capable of reproducing. In species using sexual reproduction, this implies an extra test of being able to find a mate (sexual selection). This cycle of production and evaluation may sound familiar to readers with an algorithmic background, such procedures are known as generate-and-test methods.

2.4 Evolutionary Computing: Why?

Developing automated problem solvers (that is, algorithms) is one of the central themes of mathematics and computer science. Just as engineers have always looked at Nature's solutions for inspiration, copying 'natural problem solvers' is a stream within these disciplines. When looking for the most powerful natural problem solver, there are two rather obvious candidates:

- the human brain (that created "the wheel, New York, wars and so on" [4, Chap. 23]);
- the evolutionary process (that created the human brain).

Trying to design problem solvers based on the first candidate leads to the field of neurocomputing. The second option forms a basis for evolutionary computing.

Another motivation can be identified from a technical perspective. Computerisation in the second half of the 20th century created a growing demand for problem-solving automation. The growth rate of the research and development capacity has not kept pace with these needs. Hence, the time available for thorough problem analysis and tailored algorithm design has been decreasing. A parallel trend has been the increase in the complexity of problems to

[2] Some genotype variations may not cause observable phenotype differences, but this is not relevant for the present argument.

be solved. These two trends imply an urgent need for robust algorithms with satisfactory performance. That is, there is a need for algorithms that are applicable to a wide range of problems, do not need much tailoring for specific problems, and deliver good (not necessarily optimal) solutions within acceptable time. Evolutionary algorithms do all this, and so provide an answer to the challenge of deploying automated solution methods for more and more problems, which are ever more complex, in less and less time.

A third motivation is one that can be found behind every science: human curiosity. Evolutionary processes are the subjects of scientific studies where the main objective is to understand how evolution works. From this perspective, evolutionary computing represents the possibility of performing experiments differently from traditional biology. Evolutionary processes can be simulated in a computer, where millions of generations can be executed in a matter of hours or days and repeated under various circumstances. These possibilities go far beyond studies based on excavations and fossils, or those possible *in vivo*. Naturally, the interpretation of such simulation experiments must be done very carefully. First, because we do not know whether the computer models represent the biological reality with sufficient fidelity. Second, it is unclear whether conclusions drawn in a digital medium, *in silico*, can be transferred to the carbon-based biological medium. Despite these caveats there is a strong tradition within evolutionary computing to 'play around' with evolution for the sake of understanding how it works. Application issues do not play a role here, at least not in the short term. But, of course, learning more about evolutionary processes in general can help in designing better algorithms later.

Having given three rather different reasons why people might want to use evolutionary computation, we next illustrate the power of evolutionary problem solving by a number of application examples from various areas.

A challenging optimisation task that has successfully been carried out by evolutionary algorithms is the timetabling of university classes [74, 329]. Typically, some 2000–5000 events take place during a university week, and these must each be given a day, time, and room. The first optimisation task is to reduce the number of clashes, for example, a student needing to be in two places at once, or a room being used for two lectures at the same time. Producing feasible timetables with no clashes is a hard task. In fact, it turns out that in most cases the vast majority of the space of all timetables is filled with infeasible solutions. In addition to producing feasible timetables, we also want to produce timetables that are optimised as far as the users are concerned. This optimisation task involves considering a large number of objectives that compete with each other. For example, students may wish to have no more than two classes in a row, while their lecturers may be more concerned with having whole days free for conducting research. Meanwhile, the main goal of the university management might be to make room utilisation more efficient, or to cut down the amount of movement around or between the buildings.

EC applications in industrial design optimisation can be illustrated with the case of a satellite dish holder boom. This ladder-like construction connects the satellite's body with the dish needed for communication. It is essential that this boom is stable, in particular vibration resistant, as there is no air in space that would damp vibrations that could break the whole construction. Keane et al. [245] optimised this construction using an evolutionary algorithm. The resulting structure is 20,000% (!) better than traditional shapes, but for humans it looks very strange: it exhibits no symmetry, and there is no intuitive design logic visible (Fig. 2.4). The final design looks pretty much

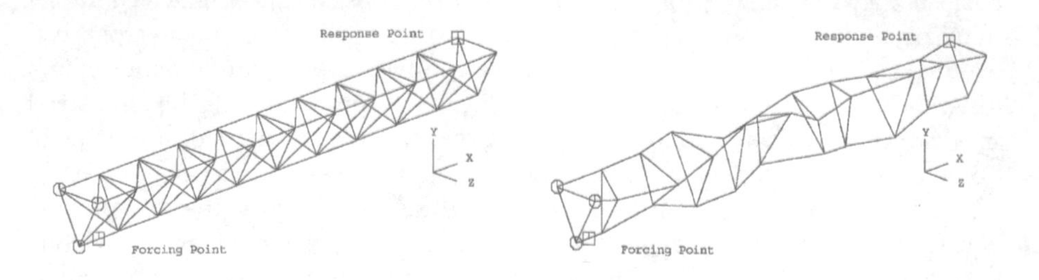

Fig. 2.4. The initial, regular design of the 3D boom (*left*) and the final design found by a genetic algorithm (*right*)

like a random drawing, and the crucial thing is this: it *is* a random drawing, drawn without intelligence, but evolving through a number of consecutive generations of improving solutions. This illustrates the power of evolution as a designer: it is not limited by conventions, aesthetic considerations, or ungrounded preferences for symmetry. On the contrary, it is purely driven by quality, and thereby it can come to solutions that lie outside of the scope of human thinking, with its implicit and unconscious limitations. It is worth mentioning that evolutionary design often goes hand-in-hand with reverse engineering. In particular, once a provably superior solution is evolved, it can be analysed and explained through the eyes of traditional engineering. This can lead to generalisable knowledge, i.e., the formulation of new laws, theories, or design principles applicable to a variety of other problems of similar type.[3]

Modelling tasks typically occur in data-rich environments. A frequently encountered situation is the presence of many examples of a certain event or phenomenon without a formal description. For instance, a bank may have one million records (profiles) of clients containing their sociogeographical data, financial overviews of their mortgages, loans, and insurances, details of their card usage, and so forth. Certainly, the bank also has information about client

[3] In the case of the satellite dish boom, it is exactly the asymmetric character that works so well. Namely, vibrations are waves that traverse the boom along the rungs. If the rungs are of different lengths then these waves meet in a different phase and cancel each other. This small theory sounds trivial, but it took the asymmetric evolved solution to come to it.

behaviour in terms of paying back loans, for instance. In this situation it is a reasonable assumption that the profile (facts and known data from the past) is related to behaviour (future events). In order to understand the repayment phenomenon, what is needed is a model relating the profile inputs to the behavioural patterns (outputs). Such a model would have predictive power, and thus would be very useful when deciding about new loan applicants. This situation forms a typical application context for the areas of machine learning and data mining. Evolutionary computing is a possible technology that has been used to solve such problems [179].

Another example of this type of modelling approach can be seen in [370], where Schulenburg and Ross use a learning classifier system to evolve sets of rules modelling the behaviour of stock market traders. As their inputs they used ten years of trading history, in the form of daily statistics such as volume of trade, current price, change in price over the last few days, whether this price is a new high (or low), and so on for a given company's stock. The evolved traders consisted of sets of `condition→action` rules. Each day the current stock market conditions were presented to the trader, triggering a rule that decided whether stock was bought or sold. Periodically a genetic algorithm is run on the set of (initially random) rules, so that well-performing ones are rewarded, and poorly performing ones are discarded. It was demonstrated that the system evolved trading agents that outperformed many well-known strategies, and varied according to the nature of the particular stock they were trading. Of particular interest, and benefit, compared to methods such as neural networks (which are also used for this kind of modelling problem in time-series forecasting), is the fact that the rule-bases of the evolved traders are easily examinable, that is to say that the models that are evolved are particularly transparent to the user.

Evolutionary computing can also be applied to simulation problems, that is, to answer what-if questions in a context where the investigated subject matter is evolving, i.e., driven by variation and selection. Evolutionary economics is an established research area, roughly based on the perception that the game and the players in the socioeconomic arena have much in common with the game of life. In common parlance, the survival of the fittest principle is also fundamental in the economic context. Evolving systems with a socioeconomic interpretation can differ from biological ones in that the behavioural rules governing the individuals play a very strong role in the system. The term agent-based computational economy is often used to emphasise this aspect [427]. Academic research in this direction is often based on a simple model called Sugarscape world [155]. This features agent-like inhabitants in a grid space, and a commodity (the sugar) that can be consumed, owned, traded, and so on by the inhabitants. There are many ways to set up system variants with an economics interpretation and conduct simulation experiments. For instance, Bäck et al. [31] investigate how artificially forced sugar redistribution (tax) and evolution interact under various circumstances. Clearly, interpretation of the outcomes of such experiments must be done very carefully, avoiding ungrounded claims on transferability of results into a real socioeconomic context.

Finally, we note that evolutionary computing experiments with a clear biological interpretation are also very interesting. Let us mention two approaches by way of illustration: trying existing biological features or trying nonexisting biological features. In the first approach, simulating a known natural phenomenon is a key issue. This may be motivated by an expectation that the natural trick will also work for algorithmic problem-solving, or by simply being willing to test whether the effects known in carbon would occur in silicon as well. Take incest as an example. A strong moral taboo against incest has existed for thousands of years, and for the last century or two there is also scientific insight supporting this: incest leads to degeneration of the population. The results in [158] show that computer-simulated evolution also benefits from incest prevention. This confirms that the negative effects of incest are inherent for evolutionary processes, independently from the medium in which they take place. The other approach to simulations with a biological flavour is the opposite of this: it implements a feature that does not exist in biology, but can be implemented in a computer. As an illustration, let us take multiparent reproduction, where more than two parents are required for mating, and offspring inherit genetic material from each of them. Eiben et al. [126, 128] have experimented a great deal with such mechanisms showing the beneficial effects under many different circumstances.

To summarise this necessarily brief introduction, evolutionary computing is a branch of computer science concerned with a class of algorithms that are broadly based on the Darwinian principles of natural selection, and that draw inspiration from molecular genetics. Over the history of the world, many species have arisen and evolved to suit different environments, all using the same biological machinery. In the same way, if we provide an evolutionary algorithm with a new environment we hope to see adaptation of the initial population in a way that better suits the environment. Typically (but not always) this environment will take the form of a problem to be solved, with feedback to the individuals representing how well the solutions they represent solve the problem, and we have provided some examples of this. However, as we have indicated, the search for optimal solutions to some problem is not the only use of evolutionary algorithms; their nature as flexible adaptive systems gives rise to applications varying from economic modelling and simulation to the study of diverse biological processes during adaptation.

For exercises and recommended reading for this chapter, please visit `www.evolutionarycomputation.org`.

3

What Is an Evolutionary Algorithm?

The most important aim of this chapter is to describe what an evolutionary algorithm (EA) is. In order to give a unifying view we present a general scheme that forms the common basis for all the different variants of evolutionary algorithms. The main components of EAs are discussed, explaining their role and related issues of terminology. This is immediately followed by two example applications to make things more concrete. We then go on to discuss general issues concerning the operation of EAs, to place them in a broader context and explain their relationship with other global optimisation techniques.

3.1 What Is an Evolutionary Algorithm?

As the history of the field suggests, there are many different variants of evolutionary algorithms. The common underlying idea behind all these techniques is the same: given a population of individuals within some environment that has limited resources, competition for those resources causes natural selection (survival of the fittest). This in turn causes a rise in the fitness of the population. Given a quality function to be maximised, we can randomly create a set of candidate solutions, i.e., elements of the function's domain. We then apply the quality function to these as an abstract fitness measure – the higher the better. On the basis of these fitness values some of the better candidates are chosen to seed the next generation. This is done by applying recombination and/or mutation to them. Recombination is an operator that is applied to two or more selected candidates (the so-called parents), producing one or more new candidates (the children). Mutation is applied to one candidate and results in one new candidate. Therefore executing the operations of recombination and mutation on the parents leads to the creation of a set of new candidates (the offspring). These have their fitness evaluated and then compete – based on their fitness (and possibly age) – with the old ones for a place in the next generation. This process can be iterated until a candidate with sufficient quality (a solution) is found or a previously set computational limit is reached.

There are two main forces that form the basis of evolutionary systems:

- Variation operators (recombination and mutation) create the necessary diversity within the population, and thereby facilitate novelty.
- Selection acts as a force increasing the mean quality of solutions in the population.

The combined application of variation and selection generally leads to improving fitness values in consecutive populations. It is easy to view this process as if evolution is optimising (or at least 'approximising') the fitness function, by approaching the optimal values closer and closer over time. An alternative view is that evolution may be seen as a process of adaptation. From this perspective, the fitness is not seen as an objective function to be optimised, but as an expression of environmental requirements. Matching these requirements more closely implies an increased viability, which is reflected in a higher number of offspring. The evolutionary process results in a population which is increasingly better adapted to the environment.

It should be noted that many components of such an evolutionary process are stochastic. For example, during selection the best individuals are not chosen deterministically, and typically even the weak individuals have some chance of becoming a parent or of surviving. During the recombination process, the choice of which pieces from the parents will be recombined is made at random. Similarly for mutation, the choice of which pieces will be changed within a candidate solution, and of the new pieces to replace them, is made randomly. The general scheme of an **evolutionary algorithm** is given in pseudocode in Fig. 3.1, and is shown as a flowchart in Fig. 3.2.

```
BEGIN
  INITIALISE population with random candidate solutions;
  EVALUATE each candidate;
  REPEAT UNTIL ( TERMINATION CONDITION is satisfied ) DO
    1 SELECT parents;
    2 RECOMBINE pairs of parents;
    3 MUTATE the resulting offspring;
    4 EVALUATE new candidates;
    5 SELECT individuals for the next generation;
  OD
END
```

Fig. 3.1. The general scheme of an evolutionary algorithm in pseudocode

It is easy to see that this scheme falls into the category of generate-and-test algorithms. The evaluation (fitness) function provides a heuristic estimate of solution quality, and the search process is driven by the variation and selection operators. Evolutionary algorithms possess a number of features that can help position them within the family of generate-and-test methods:

- EAs are population based, i.e., they process a whole collection of candidate solutions simultaneously.
- Most EAs use recombination, mixing information from two or more candidate solutions to create a new one.
- EAs are stochastic.

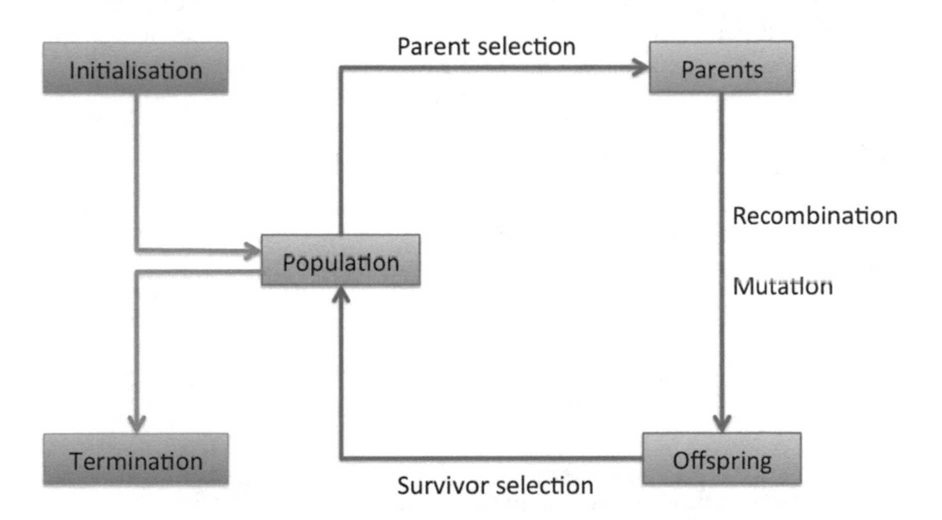

Fig. 3.2. The general scheme of an evolutionary algorithm as a flowchart

The various dialects of evolutionary computing we have mentioned previously all follow these general outlines, differing only in technical details. In particular, different streams are often characterised by the representation of a candidate solution – that is to say the data structures used to encode candidates. Typically this has the form of strings over a finite alphabet in genetic algorithms (GAs), real-valued vectors in evolution strategies (ESs), finite state machines in classical evolutionary programming (EP), and trees in genetic programming (GP). The origin of these differences is mainly historical. Technically, one representation might be preferable to others if it matches the given problem better; that is, it makes the encoding of candidate solutions easier or more natural. For instance, when solving a satisfiability problem with n logical variables, the straightforward choice is to use bit-strings of length n so that the contents of the ith bit would denote that variable i took the value $true$ (1) or $false$ (0). Hence, the appropriate EA would be a GA. To evolve a computer program that can play checkers, the parse trees of the syntactic expressions forming the programs are a natural choice to represent candidate solutions, thus a GP approach is likely. It is important to note two points. First, the recombination and mutation operators working on candidates must match the given representation. Thus, for instance, in GP the recombination operator works on trees, while in GAs it operates on strings. Second, in contrast to variation operators, the selection process only takes fitness information into account, and so it works independently from the choice of representation. Therefore differences between the selection mechanisms commonly applied in each stream are a matter of tradition rather than of technical necessity.

3.2 Components of Evolutionary Algorithms

In this section we discuss evolutionary algorithms in detail. There are a number of components, procedures, or operators that must be specified in order to define a particular EA. The most important components, indicated by italics in Fig. 3.1, are:

- representation (definition of individuals)
- evaluation function (or fitness function)
- population
- parent selection mechanism
- variation operators, recombination and mutation
- survivor selection mechanism (replacement)

To create a complete, runnable algorithm, it is necessary to specify each component and to define the initialisation procedure. If we wish the algorithm to stop at some stage[1], we must also provide a termination condition.

3.2.1 Representation (Definition of Individuals)

The first step in defining an EA is to link the 'real world' to the 'EA world', that is, to set up a bridge between the original problem context and the

[1] Note that this is not always this case. For instance, there are many examples of open-ended evolution of art on the Internet.

problem-solving space where evolution takes place. This often involves simplifying or abstracting some aspects of the real world to create a well-defined and tangible problem context within which possible solutions can exist and be evaluated, and this work is often undertaken by domain experts. The first step from the point of view of automated problem-solving is to decide how possible solutions should be specified and stored in a way that can be manipulated by a computer. We say that objects forming possible solutions within the original problem context are referred to as **phenotypes**, while their encoding, that is, the individuals within the EA, are called **genotypes**. This first design step is commonly called **representation**, as it amounts to specifying a mapping from the phenotypes onto a set of genotypes that are said to represent them. For instance, given an optimisation problem where the possible solutions are integers, the given set of integers would form the set of phenotypes. In this case one could decide to represent them by their binary code, so, for example, the value 18 would be seen as a phenotype, and 10010 as a genotype representing it. It is important to understand that the phenotype space can be very different from the genotype space, and that the whole evolutionary search takes place in the genotype space. A solution – a good phenotype – is obtained by decoding the best genotype after termination. Therefore it is desirable that the (optimal) solution to the problem at hand – a phenotype – is represented in the given genotype space. In fact, since in general we will not know in advance what that solution looks like, it is usually desirable that *all* possible feasible solutions can be represented[2].

The evolutionary computation literature contains many synonyms:

- On the side of the original problem context the terms **candidate solution**, phenotype, and **individual** are all used to denote possible solutions. The space of all possible candidate solutions is commonly called the **phenotype space**.
- On the side of the EA, the terms genotype, **chromosome**, and again individual are used to denote points in the space where the evolutionary search actually takes place. This space is often termed the **genotype space**.
- There are also many synonymous terms for the elements of individuals. A placeholder is commonly called a variable, a **locus** (plural: loci), a position, or – in a biology-oriented terminology – a **gene**. An object in such a place can be called a value or an **allele**.

It should be noted that the word 'representation' is used in two slightly different ways. Sometimes it stands for the mapping from the phenotype to the genotype space. In this sense it is synonymous with **encoding**, e.g., one could mention binary representation or binary encoding of candidate solutions. The inverse mapping from genotypes to phenotypes is usually called **decoding**, and it is necessary that the representation should be invertible so that for each

[2] In the language of generate-and-test algorithms, this means that the generator is *complete*.

genotype there is at most one corresponding phenotype. The word representation can also be used in a slightly different sense, where the emphasis is not on the mapping itself, but on the data structure of the genotype space. This interpretation is the one we use when, for example, we speak about mutation operators for binary representation.

3.2.2 Evaluation Function (Fitness Function)

The role of the **evaluation function** is to represent the requirements the population should adapt to meet. It forms the basis for selection, and so it facilitates improvements. More accurately, it defines what improvement means. From the problem-solving perspective, it represents the task to be solved in the evolutionary context. Technically, it is a function or procedure that assigns a quality measure to genotypes. Typically, this function is composed from the inverse representation (to create the corresponding phenotype) followed by a quality measure in the phenotype space. To stick with the example above, if the task is to find an integer x that maximises x^2, the fitness of the genotype 10010 could be defined by decoding its corresponding phenotype ($10010 \rightarrow 18$) and then taking its square: $18^2 = 324$.

The evaluation function is commonly called the **fitness function** in EC. This might cause a counterintuitive terminology if the original problem requires minimisation, because the term fitness is usually associated with maximisation. Mathematically, however, it is trivial to change minimisation into maximisation, and vice versa. Quite often, the original problem to be solved by an EA is an optimisation problem (treated in more technical detail in Sect. 1.1). In this case the name objective function is often used in the original problem context, and the evaluation (fitness) function can be identical to, or a simple transformation of, the given objective function.

3.2.3 Population

The role of the **population** is to hold (the representation of) possible solutions. A population is a multiset[3] of genotypes. The population forms the unit of evolution. Individuals are static objects that do not change or adapt; it is the population that does. Given a representation, defining a population may be as simple as specifying how many individuals are in it, that is, setting the population size. In some sophisticated EAs a population has an additional spatial structure, defined via a distance measure or a neighbourhood relation. This corresponds loosely to the way that real populations evolve within the context of a spatial structure given by individuals' geographical locations. In such cases the additional structure must also be defined in order to fully specify a population.

[3] A multiset is a set where multiple copies of an element are possible.

In almost all EA applications the population size is constant and does not change during the evolutionary search – this produces the limited resources need to create competition. The selection operators (parent selection and survivor selection) work at the population level. In general, they take the whole current population into account, and choices are always made relative to what is currently present. For instance, the best individual *of a given population* is chosen to seed the next generation, or the worst individual *of a given population* is chosen to be replaced by a new one. This population level activity is in contrast to variation operators, which act on one or more parent individuals.

The **diversity** of a population is a measure of the number of *different* solutions present. No single measure for diversity exists. Typically people might refer to the number of different fitness values present, the number of different phenotypes present, or the number of different genotypes. Other statistical measures such as entropy are also used. Note that the presence of only one fitness value in a population does not necessarily imply that only one phenotype is present, since many phenotypes may have the same fitness. Equally, the presence of only one phenotype does not necessarily imply only one genotype. However, if only one genotype is present then this implies only one phenotype and fitness value are present.

3.2.4 Parent Selection Mechanism

The role of **parent selection** or **mate selection** is to distinguish among individuals based on their quality, and, in particular, to allow the better individuals to become parents of the next generation. An individual is a **parent** if it has been selected to undergo variation in order to create offspring. Together with the survivor selection mechanism, parent selection is responsible for pushing quality improvements. In EC, parent selection is typically probabilistic. Thus, high-quality individuals have more chance of becoming parents than those with low quality. Nevertheless, low-quality individuals are often given a small, but positive chance; otherwise the whole search could become too greedy and the population could get stuck in a local optimum.

3.2.5 Variation Operators (Mutation and Recombination)

The role of **variation operators** is to create new individuals from old ones. In the corresponding phenotype space this amounts to generating new candidate solutions. From the generate-and-test search perspective, variation operators perform the generate step. Variation operators in EC are divided into two types based on their arity, distinguishing unary (mutation) and n-ary versions (recombination).

Mutation

A unary variation operator is commonly called **mutation**. It is applied to one genotype and delivers a (slightly) modified mutant, the **child** or **offspring**.

A mutation operator is always stochastic: its output – the child – depends on the outcomes of a series of random choices. It should be noted that not all unary operators are seen as mutation. For example, it might be tempting to use the term mutation to describe a problem-specific heuristic operator which acts systematically on one individual trying to find its weak spot and improve it by performing a small change. However, in general mutation is supposed to cause a random, unbiased change. For this reason it might be more appropriate not to call heuristic unary operators mutation. Historically, mutation has played a different role in various EC dialects. Thus, for example, in genetic programming it is often not used at all, whereas in genetic algorithms it has traditionally been seen as a background operator, providing the gene pool with 'fresh blood', and in evolutionary programming it is the only variation operator, solely responsible for the generation of new individuals.

Variation operators form the evolutionary implementation of elementary (search) steps, giving the search space its topological structure. Generating a child amounts to stepping to a new point in this space. From this perspective, mutation has a theoretical role as well: it can guarantee that the space is connected. There are theorems which state that an EA will (given sufficient time) discover the global optimum of a given problem. These often rely on this connectedness property that each genotype representing a possible solution can be reached by the variation operators [129]. The simplest way to satisfy this condition is to allow the mutation operator to jump everywhere: for example, by allowing any allele to be mutated into any other with a nonzero probability. However, many researchers feel these proofs have limited practical importance, and EA implementations often don't possess this property.

Recombination

A binary variation operator is called **recombination** or **crossover**. As the names indicate, such an operator merges information from two parent genotypes into one or two offspring genotypes. Like mutation, recombination is a stochastic operator: the choices of what parts of each parent are combined, and how this is done, depend on random drawings. Again, the role of recombination differs between EC dialects: in genetic programming it is often the only variation operator, and in genetic algorithms it is seen as the main search operator, whereas in evolutionary programming it is never used. Recombination operators with a higher arity (using more than two parents) are mathematically possible and easy to implement, but have no biological equivalent. Perhaps this is why they are not commonly used, although several studies indicate that they have positive effects on the evolution [126, 128].

The principle behind recombination is simple – by mating two individuals with different but desirable features, we can produce an offspring that combines both of those features. This principle has a strong supporting case – for millennia it has been successfully applied by plant and livestock breeders to produce species that give higher yields or have other desirable features. Evolutionary algorithms create a number of offspring by random recombination, and we hope that while some will have undesirable combinations of traits, and most may be no better or worse than their parents, some will have improved characteristics. The biology of the planet Earth, where, with *very* few exceptions, lower organisms reproduce asexually and higher organisms reproduce sexually [288, 289], suggests that recombination is the superior form of reproduction. However recombination operators in EAs are usually applied probabilistically, that is, with a nonzero chance of not being performed.

It is important to remember that variation operators are representation dependent. Thus for different representations different variation operators have to be defined. For example, if genotypes are bit-strings, then inverting a bit can be used as a mutation operator. However, if we represent possible solutions by tree-like structures another mutation operator is required.

3.2.6 Survivor Selection Mechanism (Replacement)

Similar to parent selection, the role of **survivor selection** or **environmental selection** is to distinguish among individuals based on their quality. However, it is used in a different stage of the evolutionary cycle – the survivor selection mechanism is called after the creation of the offspring from the selected parents. As mentioned in Sect. 3.2.3, in EC the population size is almost always constant. This requires a choice to be made about which individuals will be allowed in to the next generation. This decision is often based on their fitness values, favouring those with higher quality, although the concept of age is also frequently used. In contrast to parent selection, which is typically stochastic, survivor selection is often deterministic. Thus, for example, two common methods are the fitness-based method of ranking the unified multiset of parents and offspring and selecting the top segment, or the age-biased approach of selecting only from the offspring.

Survivor selection is also often called the **replacement** strategy. In many cases the two terms can be used interchangeably, but we use the name survivor selection to keep terminology consistent: steps 1 and 5 in Fig. 3.1 are both named selection, distinguished by a qualifier. Equally, if the algorithm creates surplus children (e.g., 500 offspring from a population of 100), then using the term survivor selection is clearly appropriate. On the other hand, the term "replacement" might be preferred if the number of newly-created children is small compared to the number of individuals in the population. For example, a "steady-state" algorithm might generate two children per iteration from a population of 100. In this case, survivor selection means choosing the two old individuals that are to be deleted to make space for the new ones, so it is more efficient to declare that everybody survives unless deleted and to choose whom to replace. Both strategies can of course be seen in nature, and have their proponents in EC, so in the rest of this book we will be pragmatic about this issue. We will use survivor selection in the section headers for reasons of generality and uniformity, while using replacement if it is commonly used in the literature for the given procedure we are discussing.

3.2.7 Initialisation

Initialisation is kept simple in most EA applications; the first population is seeded by randomly generated individuals. In principle, problem-specific heuristics can be used in this step, to create an initial population with higher fitness. Whether this is worth the extra computational effort, or not, very much depends on the application at hand. There are, however, some general observations concerning this question that we discuss in Sect. 3.5, and we also return to this issue in Chap. 10.

3.2.8 Termination Condition

We can distinguish two cases of a suitable **termination condition**. If the problem has a known optimal fitness level, probably coming from a known optimum of the given objective function, then in an ideal world our stopping condition would be the discovery of a solution with this fitness. If we know that our model of the real-world problem contains necessary simplifications, or may contain noise, we may accept a solution that reaches the optimal fitness to within a given precision $\epsilon > 0$. However, EAs are stochastic and mostly there are no guarantees of reaching such an optimum, so this condition might never get satisfied, and the algorithm may never stop. Therefore we must extend this condition with one that certainly stops the algorithm. The following options are commonly used for this purpose:

1. The maximally allowed CPU time elapses.
2. The total number of fitness evaluations reaches a given limit.
3. The fitness improvement remains under a threshold value for a given period of time (i.e., for a number of generations or fitness evaluations).
4. The population diversity drops under a given threshold.

Technically, the actual termination criterion in such cases is a disjunction: optimum value hit *or* condition X satisfied. If the problem does not have a known optimum, then we need no disjunction. We simply need a condition from the above list, or a similar one that is guaranteed to stop the algorithm. We will return to the issue of when to terminate an EA in Sect. 3.5.

3.3 An Evolutionary Cycle by Hand

To illustrate the working of an EA, we show the details of one selection–reproduction cycle on a simple problem after Goldberg [189], that of maximising the values of x^2 for integers in the range 0–31. To execute a full evolutionary cycle, we must make design decisions regarding the EA components representation, parent selection, recombination, mutation, and survivor selection.

For the representation we use a simple five-bit binary encoding mapping integers (phenotypes) to bit-strings (genotypes). For parent selection we use a fitness proportional mechanism, where the probability p_i that an individual i in population P is chosen to be a parent is $p_i = f(i)/\sum_{j \in P} f(j)$. Furthermore, we can decide to replace the entire population in one go by the offspring created from the selected parents. This means that our survivor selection operator is very simple: all existing individuals are removed from the population and all new individuals are added to it without comparing fitness values. This implies that we will create as many offspring as there are members in the population. Given our chosen representation, the mutation and recombination operators can be kept simple. Mutation is executed by generating a random number (from a uniform distribution over the range [0, 1]) in each bit position, and comparing it to a fixed threshold, usually called the **mutation rate**. If the random number is below that rate, the value of the gene in the corresponding position is flipped. Recombination is implemented by the classic one-point crossover. This operator is applied to two parents and produces two children by choosing a random crossover-point along the strings and swapping the bits of the parents after this point.

String no.	Initial population	x Value	Fitness $f(x) = x^2$	$Prob_i$	Expected count	Actual count
1	0 1 1 0 1	13	169	0.14	0.58	1
2	1 1 0 0 0	24	576	0.49	1.97	2
3	0 1 0 0 0	8	64	0.06	0.22	0
4	1 0 0 1 1	19	361	0.31	1.23	1
Sum			1170	1.00	4.00	4
Average			293	0.25	1.00	1
Max			576	0.49	1.97	2

Table 3.1. The x^2 example, 1: initialisation, evaluation, and parent selection

After having made the essential design decisions, we can execute a full selection–reproduction cycle. Table 3.1 shows a random initial population of four genotypes, the corresponding phenotypes, and their fitness values. The cycle then starts with selecting the parents to seed the next generation. The fourth column of Table 3.1 shows the expected number of copies of each individual after parent selection, being $f_i/\bar{f}$, where $\bar{f}$ denotes the average fitness (displayed values are rounded up). As can be seen, these numbers are not integers; rather they represent a probability distribution, and the mating pool is created by making random choices to sample from this distribution. The

String no.	Mating pool	Crossover point	Offspring after xover	x Value	Fitness $f(x) = x^2$
1	0 1 1 0 | 1	4	0 1 1 0 0	12	144
2	1 1 0 0 | 0	4	1 1 0 0 1	25	625
2	1 1 | 0 0 0	2	1 1 0 1 1	27	729
4	1 0 | 0 1 1	2	1 0 0 0 0	16	256
Sum					1754
Average					439
Max					729

Table 3.2. The x^2 example, 2: crossover and offspring evaluation

String no.	Offspring after xover	Offspring after mutation	x Value	Fitness $f(x) = x^2$
1	0 1 1 0 0	1 1 1 0 0	26	676
2	1 1 0 0 1	1 1 0 0 1	25	625
2	1 1 0 1 1	1 1 0 1 1	27	729
4	1 0 0 0 0	1 0 1 0 0	18	324
Sum				2354
Average				588.5
Max				729

Table 3.3. The x^2 example, 3: mutation and offspring evaluation

column "Actual count" stands for the number of copies in the mating pool, i.e., it shows *one* possible outcome.

Next the selected individuals are paired at random, and for each pair a random point along the string is chosen. Table 3.2 shows the results of crossover on the given mating pool for crossover points after the fourth and second genes, respectively, together with the corresponding fitness values. Mutation is applied to the offspring delivered by crossover. Once again, we show one possible outcome of the random drawings, and Table 3.3 shows the hand-made 'mutants'. In this case, the mutations shown happen to have caused positive changes in fitness, but we should emphasise that in later generations, as the number of 1's in the population rises, mutation will be *on average* (but not always) deleterious. Although manually engineered, this example shows a typical progress: the average fitness grows from 293 to 588.5, and the best fitness in the population from 576 to 729 after crossover and mutation.

3.4 Example Applications

3.4.1 The Eight-Queens Problem

This is the problem of placing eight queens on a regular 8×8 chessboard so that no two of them can check each other. This problem can be naturally generalised, yielding the N-queens problem described in Sect. 1.3. There are many classical artificial intelligence approaches to this problem, which work in a constructive, or incremental, fashion. They start by placing one queen, and after having placed n queens, they attempt to place the $(n+1)$th in a feasible position where the new queen does not check any others. Typically some sort of backtracking mechanism is applied; if there is no feasible position for the $(n+1)$th queen, the nth is moved to another position.

An evolutionary approach to this problem is drastically different in that it is not incremental. Our candidate solutions are complete (rather than partial) board configurations, which specify the positions of all eight queens. The phenotype space P is the set of all such configurations. Clearly, most elements of this space are infeasible, violating the condition of nonchecking queens. The quality $q(p)$ of any phenotype $p \in P$ can be simply quantified by the number of checking queen pairs. The lower this measure, the better a phenotype (board configuration), and a zero value, $q(p) = 0$, indicates a good solution. From this observation we can formulate a suitable objective function to be minimised, with a known optimal value. Even though we have not defined genotypes at this point, we can state that the fitness (to be maximised) of a genotype g that represents phenotype p is some inverse of $q(p)$. There are many possible ways of specifying what kind of inverse we wish to use here. For instance, $1/q(p)$ is an easy option, but has the disadvantage that attempting division by zero is a problem for many computing systems. We could circumvent this by watching for $q(p) = 0$ and saying that when this occurs we have a solution, or by adding a small value ϵ, i.e., $1/(q(p) + \epsilon)$. Other options are to use $-q(p)$ or $M \quad q(p)$, where M is a sufficiently large number to make all fitness values positive, e.g., $M \geq max\{q(p) \mid p \in P\}$. This fitness function inherits the property of q that it has a known optimum M.

To design an EA to search the space P we need to define a representation of phenotypes from P. The most straightforward idea is to use a matrix representation of elements of P directly as genotypes, meaning that we must design variation operators for these matrices. In this example, however, we define a more clever representation as follows. A genotype, or chromosome, is a permutation of the numbers $1, \ldots, 8$, and a given $g = \langle i_1, \ldots, i_8 \rangle$ denotes the (unique) board configuration, where the nth column contains exactly one queen placed on the i_nth row. For instance, the permutation $g = \langle 1, \ldots, 8 \rangle$ represents a board where the queens are placed along the main diagonal. The genotype space G is now the set of all permutations of $1, \ldots, 8$ and we also have defined a mapping $F : G \rightarrow P$.

It is easy to see that by using such chromosomes we restrict the search to board configurations where horizontal constraint violations (two queens on the same row) and vertical constraint violations (two queens on the same column) do not occur. In other words, the representation guarantees half of the requirements of a solution – what remains to be minimised is the number of diagonal constraint violations. From a formal perspective we have chosen a representation that is not surjective since only part of P can be obtained by decoding elements of G. While in general this could carry the danger of missing solutions in P, in our present example this is not the case, since we know a priori that those phenotypes from $P \setminus F(G)$ can never be solutions.

The next step is to define suitable variation operators (mutation and crossover) for our representation, i.e., to work on genotypes that are permutations. The crucial feature of a suitable operator is that it does not lead out of the space G. In common parlance, the offspring of permutations must themselves be permutations. Later, in Sects. 4.5.1 and 4.5.2, we will discuss such operators in great detail. Here we only describe one suitable mutation and one crossover operator for the purpose of illustration. For mutation we can use an operator that randomly selects two positions in a given chromosome, and swaps the values found in those positions. A good crossover for permutations is less obvious, but the mechanism outlined in Fig. 3.3 will create two child permutations from two parents.

1. Select a random position, the crossover point, $i \in \{1, \ldots, 7\}$
2. Cut both parents into two segments at this position
3. Copy the first segment of parent 1 into child 1 and the first segment of parent 2 into child 2
4. Scan parent 2 from left to right and fill the second segment of child 1 with values from parent 2, skipping those that it already contains
5. Do the same for parent 1 and child 2

Fig. 3.3. 'Cut-and-crossfill' crossover

The important thing about these variation operators is that mutation causes a small undirected change, and crossover creates children that inherit genetic material from both parents. It should be noted though that there can be large performance differences between operators, e.g., an EA using mutation A might find a solution quickly, whereas one using mutation B might never find a solution. The operators we sketch here are not necessarily *efficient*; they merely serve as examples of operators that are *applicable* to the given representation.

The next step in setting up an EA is to decide upon the selection and population update mechanisms. We will choose a simple scheme for managing the population. In each evolutionary cycle we will select two parents, producing two children, and the new population of size n will contain the best n of the resulting $n + 2$ individuals (the old population plus the two new ones).

Parent selection (step 1 in Fig. 3.1) will be done by choosing five individuals randomly from the population and taking the best two as parents. This ensures a bias towards using parents with relatively high fitness. Survivor selection (step 5 in Fig. 3.1) checks which old individuals should be deleted to make place for the new ones – provided the new ones are better. Following the naming convention discussed from Sect. 3.2.6 we define a replacement strategy. The strategy we will use merges the population and offspring, then ranks them according to fitness, and deletes the worst two.

To obtain a full specification we can decide to fill the initial population with randomly generated permutations, and to terminate the search when we find a solution, or when 10,000 fitness evaluations have elapsed, whichever happens sooner. Furthermore we can decide to use a population size of 100, and to use the variation operators with a certain frequency. For instance, we always apply crossover to the two selected parents and in 80% of the cases apply mutation to the offspring. Putting this all together, we obtain an EA as summarised in Table 3.4.

Representation	Permutations
Recombination	'Cut-and-crossfill' crossover
Recombination probability	100%
Mutation	Swap
Mutation probability	80%
Parent selection	Best 2 out of random 5
Survival selection	Replace worst
Population size	100
Number of offspring	2
Initialisation	Random
Termination condition	Solution or 10,000 fitness evaluations

Table 3.4. Description of the EA for the eight-queens problem

3.4.2 The Knapsack Problem

The 0–1 knapsack problem, a generalisation of many industrial problems, can be briefly described as follows. We are given a set of n items, each of which has attached to it some value v_i, and some cost c_i. The task is to select a subset of those items that maximises the sum of the values, while keeping the summed cost within some capacity C_{max}. Thus, for example, when packing a backpack for a round-the-world trip, we must balance likely utility of the items against the fact that we have a limited volume (the items chosen must fit in one bag), and weight (airlines impose fees for luggage over a given weight).

It is a natural idea to represent candidate solutions for this problem as binary strings of length n, where a 1 in a given position indicates that an item is included and a 0 that it is omitted. The corresponding genotype space G is the set of all such strings with size 2^n, which increases exponentially with the number of items considered. Using this G, we fix the representation in the sense of data structure, and next we need to define the mapping from genotypes to phenotypes.

The first representation (in the sense of a mapping) that we consider takes the phenotype space P and the genotype space to be identical. The quality of a given solution p, represented by a binary genotype g, is thus determined by summing the values of the included items, i.e., $q(p) = \sum_{i=1}^{n} v_i \cdot g_i$. However, this simple representation leads us to some immediate problems. By using a one-to-one mapping between the genotype space G and the phenotype space P, individual genotypes may correspond to invalid solutions that have an associated cost greater than the capacity, i.e., $\sum_{i=1}^{n} c_i \cdot g_i > C_{max}$. This issue is typical of a class of problems that we return to in Chap. 13, and a number of mechanisms have been proposed for dealing with it.

The second representation that we outline here solves this problem by employing a decoder function, that breaks the one-to-one correspondence between the genotype space G and the solution space P. In essence, our genotype representation remains the same, but when creating a solution we read from left to right along the binary string, and keep a running tally of the cost of included items. When we encounter a value 1, we first check to see whether including the item would break our capacity constraint. In other words, rather than interpreting a value 1 as meaning *include this item*, we interpret it as meaning *include this item IF it does not take us over the cost constraint*. The effect of this scheme is to make the mapping from genotype to phenotype space many-to-one, since once the capacity has been reached, the values of all bits to the right of the current position are irrelevant, as no more items will be added to the solution. Furthermore, this mapping ensures that all binary strings represent valid solutions with a unique fitness (to be maximised).

Having decided on a fixed-length binary representation, we can now choose off-the-shelf variation operators from the GA literature, because the bit-string representation is 'standard' there. A suitable (but not necessarily optimal) recombination operator is the so-called one-point crossover, where we align two parents and pick a random point along their length. The two offspring are created by exchanging the tails of the parents at that point. We will apply this with 70% probability, i.e., for each pair of parents there is a 70% chance that we will create two offspring by crossover and 30% that the children will be just copies of the parents. A suitable mutation operator is so-called bit-flipping: in each position we invert the value with a small probability $p_m \in [0, 1)$.

In this case we will create the same number of offspring as we have members in our initial population. As noted above, we create two offspring from each two parents, so we will select that many parents and pair them randomly. We will use a tournament for selecting the parents, where each time we pick two members of the population at random (with replacement), and the one with the highest value $q(p)$ wins the tournament and becomes a parent. We will institute a generational scheme for survivor selection, i.e., all of the population in each iteration are discarded and replaced by their offspring.

Finally, we should consider initialisation (which we will do by random choice of 0 and 1 in each position of our initial population), and termination. In this case, we do not know the maximum value that we can achieve, so we will run our algorithm until no improvement in the fitness of the best member of the population has been observed for 25 generations.

We have already defined our crossover probability as 0.7; we will work with a population size of 500 and a mutation rate of $p_m = 1/n$, i.e., that will *on average* change one value in every offspring. Our evolutionary algorithm to tackle this problem can be specified as below in Table 3.5.

Representation	Binary strings of length n
Recombination	One-point crossover
Recombination probability	70%
Mutation	Each value inverted with independent probability p_m
Mutation probability p_m	$1/n$
Parent selection	Best out of random 2
Survival selection	Generational
Population size	500
Number of offspring	500
Initialisation	Random
Termination condition	No improvement in last 25 generations

Table 3.5. Description of the EA for the knapsack problem

3.5 The Operation of an Evolutionary Algorithm

Evolutionary algorithms have some rather general properties concerning how they work. To illustrate how an EA typically works, we will assume a one-dimensional objective function to be maximised. Figure 3.4 shows three stages of the evolutionary search, showing how the individuals might typically be distributed in the beginning, somewhere halfway, and at the end of the evolution. In the first stage directly after initialisation, the individuals are randomly spread over the whole search space (Fig. 3.4, left). After only a few generations this distribution changes: because of selection and variation operators the population abandons low-fitness regions and starts to climb the hills (Fig. 3.4, middle). Yet later (close to the end of the search, if the termination condition is set appropriately), the whole population is concentrated around a few peaks, some of which may be suboptimal. In principle it is possible that the population might climb the wrong hill, leaving all of the individuals positioned around a local but not global optimum. Although there is no universally accepted rigorous definition of the terms exploration and exploitation, these notions are often used to categorize distinct phases of the search process. Roughly speaking, **exploration** is the generation of new individuals in as-yet untested regions of the search space, while **exploitation** means the concentration of the search in the vicinity of known good solutions. Evolutionary search processes are often referred to in terms of a trade-off between exploration and exploitation. Too much of the former can lead to inefficient search, and too much of the latter can lead to a propensity to focus the search too quickly (see [142] for a good discussion of these issues). **Premature convergence** is the well-known effect of losing population diversity too quickly, and getting trapped in a local optimum. This danger is generally present in evolutionary algorithms, and techniques to prevent it are discussed in Chap. 5.

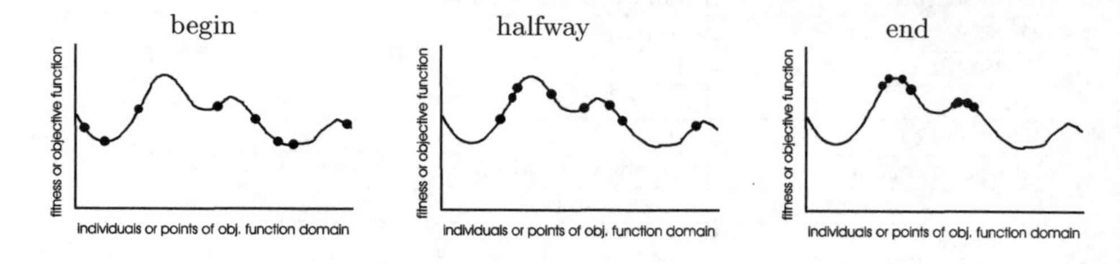

Fig. 3.4. Typical progress of an EA illustrated in terms of population distribution. For each point x in the search space y shows the corresponding fitness value.

The other effect we want to illustrate is the **anytime behaviour** of EAs by plotting the development of the population's best fitness value over time (Fig. 3.5). This curve shows rapid progress in the beginning and flattening out later on. This is typical for many algorithms that work by iterative improvements to the initial solution(s). The name 'anytime' comes from the property that the search can be stopped at any time, and the algorithm will have some solution, even if it is suboptimal. Based on this anytime curve we can

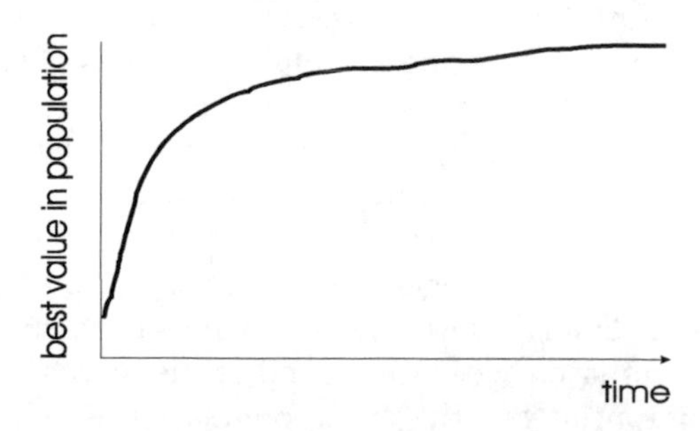

Fig. 3.5. Typical progress of an EA illustrated in terms of development over time of the highest fitness in the population

make some general observations concerning initialisation and the termination condition for EAs. In Sect. 3.2.7 we questioned whether it is worth putting extra computational effort into applying intelligent heuristics to seed the initial population with better-than-random individuals. In general, it could be said that that the typical progress curve of an evolutionary process makes it unnecessary. This is illustrated in Fig. 3.6. As the figure indicates, using

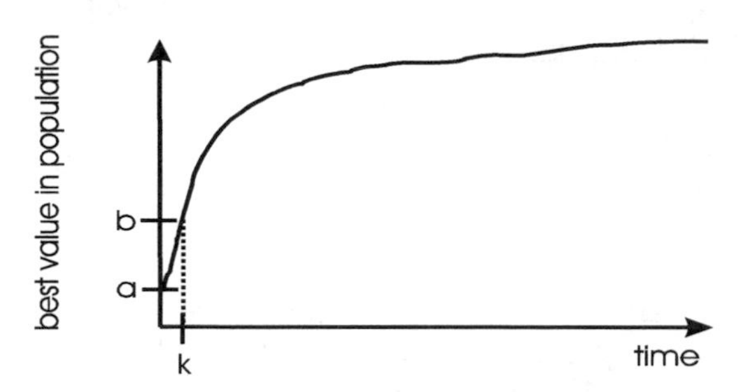

Fig. 3.6. Illustration of why heuristic initialisation might not be worth additional effort. Level a shows the best fitness in a randomly initialised population; level b belongs to heuristic initialisation

heuristic initialisation can start the evolutionary search with a better population. However, typically a few (k in the figure) generations are enough to reach this level, making the extra effort questionable. In Chap. 10 we will return to this issue.

The anytime behaviour also gives some general indications regarding the choice of termination conditions for EAs. In Fig. 3.7 we divide the run into two equally long sections. As the figure indicates, the progress in terms of fitness increase in the first half of the run (X) is significantly greater than in the second half (Y). This suggests that it might not be worth allowing very long runs. In other words, because of frequently observed anytime behaviour of EAs, we might surmise that effort spent after a certain time (number of fitness evaluations) is unlikely to result in better solution quality.

We close this review of EA behaviour by looking at EA performance from a global perspective. That is, rather than observing one run of the algorithm, we consider the performance of EAs for a wide range of problems. Fig. 3.8 shows the 1980s view after Goldberg [189]. What the figure indicates is that EAs show a roughly evenly good performance over a wide range of problems. This performance pattern can be compared to random search and to algorithms tailored to a specific problem type. EAs are suggested to clearly outperform random search. In contrast, a problem-tailored algorithm performs much better than an EA, but only on the type of problem for which it was designed. As we move away from this problem type to different problems, the problem-specific algorithm quickly loses performance. In this sense, EAs and problem-specific algorithms form two opposing extremes. This perception played an important

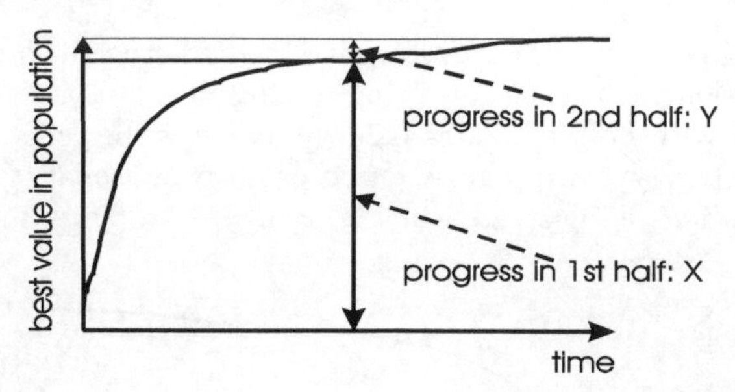

Fig. 3.7. Why long runs might not be worth performing. X shows the fitness increase in the first half of the run, while Y belongs to the second half

role in positioning EAs and stressing the difference between evolutionary and random search, but it gradually changed in the 1990s based on new insights from practice as well as from theory. The contemporary view acknowledges the possibility of combining the two extremes into a hybrid algorithm. This issue is treated in detail in Chap. 10, where we also present the revised version of Fig. 3.8. As for theoretical considerations, the No Free Lunch theorem has shown that (under some conditions) no black-box algorithm can outperform random walk when averaged over 'all' problems [467]. That is, showing the EA line always above that of random search is fundamentally incorrect. This is discussed further in Chap. 16.

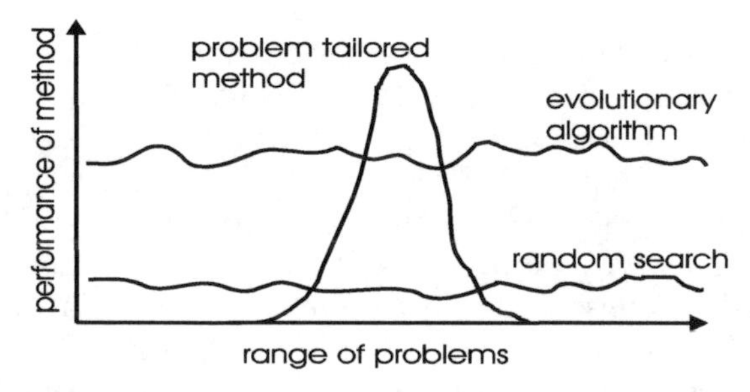

Fig. 3.8. 1980s view of EA performance after Goldberg [189]

3.6 Natural Versus Artificial Evolution

From the perspective of the underlying substrate, the emergence of evolutionary computation can be considered as a major transition of the evolutionary principles from wetware, the realm of biology, to software, the realm of computers. This was made possible by using computers as instruments for creating digital worlds that are very flexible and much more controllable than the physical reality we live in. Together with the increased understanding of the genetic mechanisms behind evolution this brought about the opportunity to become active masters of evolutionary processes that are fully designed and executed by human experimenters from above.

It could be argued that evolutionary algorithms are not faithful models of natural evolution. However, they certainly are a form of evolution. As phrased by Dennett [116]: If you have variation, heredity, and selection, then you must get evolution. In Table 3.6 we compare natural evolution and artificial evolution as used in contemporary evolutionary algorithms.

	Natural evolution	Artificial evolution
Fitness	Observed quantity: *a posteriori* effect of selection ('in the eye of the observer').	Predefined *a priori* quantity that drives selection.
Selection	Complex multifactor force based on environmental conditions, other individuals of the same species and other species (e.g., predators). Viability is tested continually; reproducibility is tested at discrete times.	Randomized operator with selection probabilities based on given fitness values. Parent selection and survivor selection both happen at discrete times.
Genotype-phenotype mapping	Highly complex biochemical process influenced by the environment.	Relatively simple mathematical transformation or parameterised procedure.
Variation	Offspring created from one (asexual reproduction) or two parents (sexual reproduction).	Offspring may be generated from one, two, or many parents.
Execution	Parallel, decentralized execution; birth and death events are not synchronised.	Typically centralized with synchronised birth and death.
Population	Spatial embedding implies structured populations. Population size varies according to the relative number of death and birth events.	Typically unstructured and panmictic (all individuals are potential partners). Population size is kept constant by synchronising time and number of birth and death events.

Table 3.6. Differences between natural and artificial evolution

3.7 Evolutionary Computing, Global Optimisation, and Other Search Algorithms

In Chap. 2 we noted that evolutionary algorithms are often used for problem optimisation. Of course EAs are not the only optimisation technique known, so in this section we explain where EAs fall into the general class of optimisation methods, and why they are of increasing interest.

In an ideal world, we would possess the technology and algorithms that could provide a provably optimal solution to any problem that we could suitably pose to the system. In fact such algorithms do exist: an exhaustive enumeration of all of the possible solutions to a problem is clearly such an algorithm. Moreover, for many problems that can be expressed in a suitably mathematical formulation, much faster, exact techniques such as branch and bound search are well known. However, despite the rapid progress in computing technology, and even if there is no halt to Moore's Law, all too often the types of problems posed by users exceed in their demands the capacity of technology to answer them.

Decades of computer science research have taught us that many real-world problems can be reduced in their essence to well-known abstract forms, for which the number of potential solutions grows very quickly with the number of variables considered. For example, many problems in transportation can be reduced to the well-known travelling salesperson problem (TSP): given a list of destinations, construct the shortest tour that visits each destination exactly once. If we have n destinations, with symmetric distances between them, the number of possible tours is $n!/2 = n \cdot (n-1) \cdot (n-2) \cdot \ldots \cdot 3$, which is exponential in n. For some of these abstract problems exact methods are known whose time complexity scales linearly (or at least polynomially) with the number of variables (see [212] for an overview). However, it is widely accepted that for many types of problems encountered, no such algorithms exist — as was discussed in Sect. 1.4. Thus, despite the increase in computing power, beyond a certain size of problem we must abandon the search for provably optimal solutions, and look to other methods for finding good solutions.

The term **global optimisation** refers to the process of attempting to find the solution with the optimal value for some fitness function. In mathematical terminology, we are trying to find the solution x^* out of a set of possible solutions S, such that $x \neq x^* \Rightarrow f(x^*) \geq f(x)\ \forall x \in S$. Here we have assumed a maximisation problem – the inequality is simply reversed for minimisation.

As noted above, a number of *deterministic* algorithms exist that, if allowed to run to completion, are guaranteed to find x^*. The simplest example is, of course, complete enumeration of all the solutions in S, which can take an exponentially long time as the number of variables increases. A variety of other techniques, collectively known as box decomposition, are based on ordering the elements of S into some kind of tree, and then reasoning about the quality of solutions in each branch in order to decide whether to investigate its elements. Although methods such as branch and bound can sometimes make very fast progress, in the worst case (caused by searching in a suboptimal order) the time complexity of the algorithms is still the same as complete enumeration.

Another class of search methods is known as *heuristics*. These may be thought of as sets of rules for deciding which potential solution out of S should next be generated and tested. For some *randomised* heuristics, such as **simulated annealing** [2, 250] and certain variants of EAs, convergence proofs do in fact exist, i.e., they are guaranteed to find x^*. Unfortunately these algorithms are fairly weak, in the sense that they will not identify x^* as being globally optimal, rather as simply the best solution seen so far.

An important class of heuristics is based on the idea of using operators that impose some kind of structure onto the elements of S, such that each point x has associated with it a set of neighbours $N(x)$. In Fig. 2.2 the variables (traits) x and y were taken to be real-valued, which imposes a natural structure on S. The reader should note that for those types of problem where each variable takes one of a finite set of values (so-called **combinatorial optimisation**), there are many possible neighbourhood structures. As an example of how the landscape 'seen' by a local search algorithm depends on its neighbourhood structure, the reader might wish to consider what a chessboard would look like if we reordered it, so that squares that are possible next moves for the knight piece were adjacent to each other. Thus points which are locally optimal (fitter than all their neighbours) in the landscape induced by one neighbourhood structure may not be for another. However, by its definition, the **global optimum** x^* will always be fitter than all of its neighbours *under any neighbourhood structure*.

So-called **local search** algorithms [2] and their many variants work by taking a starting solution x, and then searching the candidate solutions in $N(x)$ for one x' that performs better than x. If such a solution exists, then this is accepted as the new incumbent solution, and the search proceeds by examining the candidate solutions in $N(x')$. This process will eventually lead to the identification of a **local optimum**: a solution that is superior to all those in its neighbourhood. Such algorithms (often referred to as **hill climbers** for maximisation problems) have been well studied over the decades. They have the advantage that they are often quick to identify a good solution to the problem, which is sometimes all that is required in practical applications. However, the downside is that problems will frequently exhibit numerous local optima, some of which may be significantly worse than the global optimum, and no guarantees can be offered for the quality of solution found.

A number of methods have been proposed to get around this problem by changing the search landscape, either by changing the neighbourhood structure (e.g., variable neighbourhood search [208]), or by temporarily assigning low fitness to already-seen good solutions (e.g., Tabu search [186]). However the theoretical basis behind these algorithms is still very much in gestation.

There are a number of features of EAs that distinguish them from local search algorithms, relating principally to their use of a population. The population provides the algorithm with a means of defining a nonuniform probability distribution function (p.d.f.) governing the generation of new points from S. This p.d.f. reflects possible interactions between points in S which are currently represented in the population. The interactions arise from the recombination of partial solutions from two or more members of the population (parents). This potentially complex p.d.f. contrasts with the globally uniform distribution of blind random search, and the locally uniform distribution used by many other stochastic algorithms such as simulated annealing and various hill-climbing algorithms.

The ability of EAs to maintain a diverse set of points provides not only a means of escaping from local optima, but also a means of coping with large and discontinuous search spaces. In addition, as will be seen in later chapters, if several copies of a solution can be generated, evaluated, and maintained in the population, this provides a natural and robust way of dealing with problems where there is noise or uncertainty associated with the assignment of a fitness score to a candidate solution.

For exercises and recommended reading for this chapter, please visit `www.evolutionarycomputation.org`.

4

Representation, Mutation, and Recombination

As explained in Chapt. 3, there are two fundamental forces that form the basis of evolutionary systems: variation and selection. In this chapter we discuss the EA components behind the first one. Since variation operators work at the equivalent of the genetic level, that is to say they work on the representation of solutions, rather than on solutions themselves, this chapter is subdivided into sections that deal with different ways in which solutions can be represented and varied within the overall search algorithm.

4.1 Representation and the Roles of Variation Operators

The first stage of building any evolutionary algorithm is to decide on a genetic **representation** of a candidate solution to the problem. This involves defining the genotype and the mapping from genotype to phenotype. When choosing a representation, it is important to choose the right representation for the problem being solved. In many cases there will be a range of options, and getting the representation right is one of the most difficult parts of designing a good evolutionary algorithm. Often this only comes with practice and a good knowledge of the application domain. In the following sections, we look more closely at some commonly used representations, and the genetic operators that might be applied to them. It is important to stress, however, that while the representations described here are commonly used, they might not be the best representations for your application. Equally, although we present the representations and their associate operators separately, it frequently turns out in practice that using mixed representations is a more natural and suitable way of describing and manipulating a solution than trying to shoehorn different aspects of a problem into a common form.

Mutation is the generic name given to those variation operators that use only one parent and create one child by applying some kind of randomised change to the representation (genotype). The form taken depends on the choice of encoding used, as does the meaning of the associated parameter, which is often introduced to regulate the intensity or magnitude of mutation. Depending on the given implementation, this can be mutation probability, mutation rate, mutation step size, etc. In the descriptions below we concentrate on the choice of operators rather than of parameters. However, the latter can make a significant difference in the behaviour of the evolutionary algorithm, and this is discussed in more depth in Chap. 7.

Recombination, the process whereby a new individual solution is created from the information contained within two (or more) parent solutions, is considered by many to be one of the most important features in evolutionary algorithms. A lot of research activity has focused on it as the primary mechanism for creating diversity, with mutation considered as a background search operator. However, different strands of EC historically emphasised different variation operators, and as these came together under the umbrella of evolutionary algorithms, this emphasis prompted a great deal of debate. Regardless of the merits of different viewpoints, the ability to combine partial solutions via recombination is certainly one of the features that most distinguishes EAs from other global optimisation algorithms.

Although the term recombination has come to be used for the more general case, early authors used the term **crossover**, motivated by the biological analogy to meiosis (see Sect. 2.3.2). Therefore we will occasionally use the terms interchangeably, although crossover tends to refer to the most common two-parent case. Recombination operators are usually applied probabilistically according to a **crossover rate** p_c. Usually two parents are selected and two offspring are created via recombination of the two parents with probability p_c; or by simply copying the parents, with probability $1 - p_c$.

Distinguishing variation operators by their arity a makes it a straightforward idea to go beyond the usual $a = 1$ (mutation) and $a = 2$ (crossover). The resulting **multiparent recombination** operators for $a = 3, 4, \ldots$ are simple to define and implement. This provides the opportunity to experiment with evolutionary processes using reproduction schemes that do not exist in biology. From the technical point of view this offers a tool for amplifying the effects of recombination. Although such operators are not widely used in EC, there are many examples that have been proposed during the development of the field, even as early as 1966 [67], see [126, 128] for an overview, and Sect. 6.6 for a description of how this idea is applied in differential evolution. These operators can be categorised by the basic mechanism used for combining the information of the parent individuals. This mechanism can be:

- based on allele frequencies, e.g., p-sexual voting [311] generalising uniform crossover;
- based on segmentation and recombination of the parents, e.g., the diagonal crossover in [139]; generalising n-point crossover
- based on numerical operations on real-valued alleles, e.g., the centre of mass crossover [434], generalising arithmetic recombination operators.

In general, it cannot be claimed that increasing the arity of recombination has a positive effect on the performance of an EA – this depends very much on the type of recombination and the problem at hand. However, systematic studies on landscapes with tuneable ruggedness [143] and a large number of experimental investigations on various problems clearly show that using more than two parents can accelerate evolutionary search and be advantageous in many cases.

4.2 Binary Representation

The first representation we look at is one of the simplest – the binary one used in Sect. 3.3. This is one of the earliest representations, and historically many genetic algorithms (GAs) have (mistakenly) used this representation almost independently of the problem they were trying to solve. Here the genotype consists simply of a string of binary digits – a bit-string.

For a particular application we have to decide how long the string should be, and how we will interpret it to produce a phenotype. In choosing the genotype–phenotype mapping for a specific problem, one has to make sure that the encoding allows that all possible bit strings denote a valid solution to the given problem[1] and that, vice versa, all possible solutions can be represented.

For some problems, particularly those concerning Boolean decision variables, the genotype–phenotype mapping is natural. One example is the knapsack problem described in Sect. 3.4.2, where for each possible item a Boolean decision was evolved, denoting whether it was included in the final solution. Frequently bit-strings are used to encode other nonbinary information. For example, we might interpret a bit-string of length 80 as 10 integers, each encoded as 8-bit integers (allowing for 256 possible values), or five 16-bit real numbers. Using bit-strings to encode nonbinary information is usually a mistake, and better results can be obtained by using the integer or real-valued representations directly.

One of the problems of coding numbers in binary is that different bits have different significance, and so the effect of a single bit mutation is very variable. Using standard binary code has the disadvantage that the Hamming distance between two consecutive integers is often not equal to one. If the goal is to evolve an integer number, you would like to have equal probabilities of changing a 7 into an 8 or a 6. However, changing 0111 to 1000 requires four bit-flips while changing it to 0110 takes just one. Thus with a mutation operator that randomly, and independently, changes each allele value with probability $p_m < 0.5$, the probability of changing 7 to 8 is much less than changing 7 to 6. This can be helped by using **Gray coding**, a variation on the way that integers are mapped on bit strings where consecutive integers always have Hamming distance one.

[1] In practice this restriction to validity in not always possible; see Chap. 13 for a more complete discussion of this issue.

4.2.1 Mutation for Binary Representation

Although a few other schemes have been occasionally used, the most common mutation operator for binary encodings considers each gene separately and allows each bit to flip (i.e., from 1 to 0 or 0 to 1) with a small probability p_m. The actual number of values changed is thus not fixed, but depends on the sequence of random numbers drawn, so for an encoding of length L, on average $L \cdot p_m$ values will be changed. In Fig. 4.1 this is illustrated for the case where the third, fourth, and eighth random values generated are less than the bitwise mutation rate p_m.

Fig. 4.1. Bitwise mutation for binary encodings

A number of studies and recommendations have been made for the choice of suitable values for the bitwise mutation rate p_m. Most binary coded GAs use mutation rates in a range such that on average between one gene per generation and one gene per offspring is mutated. However, it is worth noting at the outset that the most suitable choice to use depends on the desired outcome. For example, does the application require a population in which *all* members have high fitness, or simply that *one* highly fit individual is found? The former suggests a lower mutation rate, less likely to disrupt good solutions. In the latter case one might choose a higher mutation rate if the potential benefits of ensuring good coverage of the search space outweighed the cost of disrupting copies of good solutions[2].

4.2.2 Recombination for Binary Representation

Three standard forms of recombination are generally used for binary representations. They all start from two parents and create two children, although all of these have been extended to the more general case where a number of parents may be used [152], and there are also situations in which only one of the offspring might be considered (Sect. 5.1).

One-Point Crossover One-point crossover was the original recombination operator proposed in [220] and examined in [102]. It works by choosing a

[2] In fact this example illustrates that the algorithm's parameters cannot be chosen independently: in the second case we might couple higher mutation rates with a more aggressive selection policy to ensure the best solutions were not lost.

random number r in the range $[1, l-1]$ (with l the length of the encoding), and then splitting both parents at this point and creating the two children by exchanging the tails (Fig. 4.2, top). Note that by using the range $[1, l-1]$ the crossover point is prevented from falling before the first position ($r = 0$) or after the last position ($r = l$).

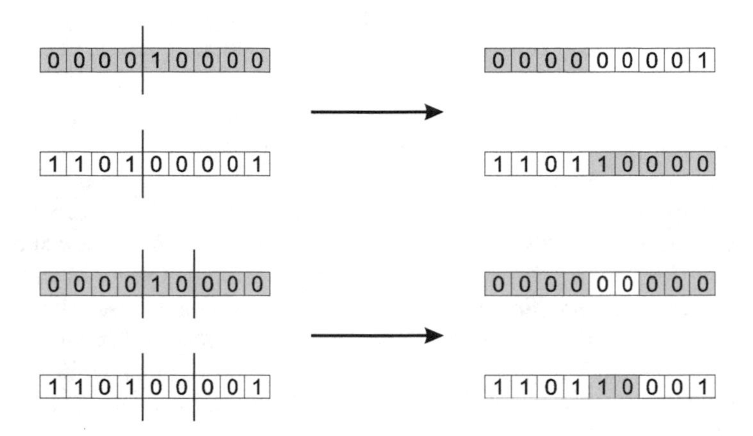

Fig. 4.2. One-point crossover (top) and n-point crossover with $n = 2$ (bottom)

n-Point Crossover One-point crossover can easily be generalised to n-point crossover, where the chromosome is broken into more than two segments of contiguous genes, and the offspring are created by taking alternative segments from the parents. In practice this means choosing n random crossover points in $[1, l-1]$, which is illustrated in Fig. 4.2 (bottom) for $n = 2$.

Uniform Crossover The previous two operators worked by dividing the parents into a number of sections of contiguous genes and reassembling them to produce offspring. In contrast to this, uniform crossover [422] works by treating each gene independently and making a random choice as to which parent it should be inherited from. This is implemented by generating a string of l random variables from a uniform distribution over [0,1]. In each position, if the value is below a parameter p (usually 0.5), the gene is inherited from the first parent; otherwise from the second. The second offspring is created using the inverse mapping. This is illustrated in Fig. 4.3.

In our discussion so far, we have suggested that in the absence of prior information, recombination worked by randomly mixing parts of the parents. However, as Fig. 4.2 illustrates, n-point crossover has an inherent bias in that it tends to keep together genes that are located close to each other in

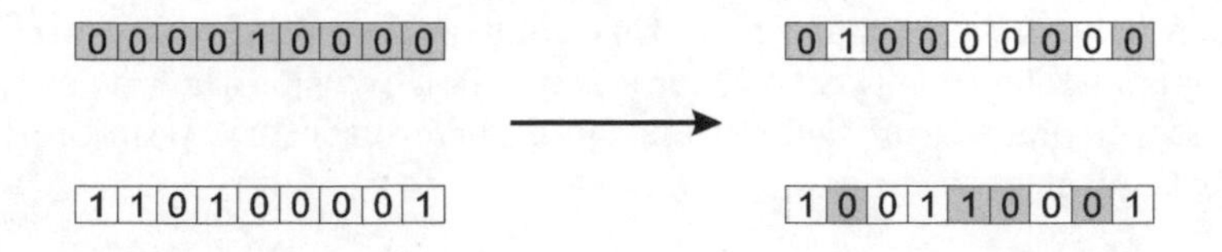

Fig. 4.3. Uniform crossover. The array [0.3, 0.6, 0.1, 0.4, 0.8, 0.7, 0.3, 0.5, 0.3] of random numbers and $p = 0.5$ were used to decide inheritance for this example.

the representation. Furthermore, when n is odd (e.g., one-point crossover), there is a strong bias against keeping together combinations of genes that are located at opposite ends of the representation. These effects are known as **positional bias** and have been extensively studied from both a theoretical and experimental perspective [157, 412] (see Sect. 16.1 for more details). In contrast, uniform crossover does not exhibit any positional bias. However, unlike n-point crossover, uniform crossover does have a strong tendency towards transmitting 50% of the genes from each parent and against transmitting an offspring a large number of coadapted genes from one parent. This is known as **distributional bias**.

The general nature of these algorithms (and the No Free Lunch theorem [467], Sect. 16.10) make it impossible to state that one or the other of these operators performs best on any given problem. Nevertheless, an understanding of the types of bias exhibited by different recombination operators can be invaluable when designing an algorithm for a particular problem, particularly if there are known patterns or dependencies in the chosen representation that can be exploited. To use the knapsack problem as an example, it might make sense to use an operator that is likely to keep together the decisions for the first few heaviest items. If the items are ordered by weight (cost) in our representation, then we could make this more likely by using n-point crossover with its positional bias. However, if we used a random ordering this might actually make it less likely that co-adapted values for certain decisions were transmitted together, so we might prefer uniform crossover.

4.3 Integer Representation

As we hinted in the previous section, binary representations are not always the most suitable if our problem more naturally maps onto a representation where different genes can take one of a set of values. One obvious example of when this might occur is the problem of finding the optimal values for a set of variables that all take integer values. These values might be unrestricted (i.e., any integer value is permissible), or might be restricted to a finite set: for example, if we are trying to evolve a path on a square grid, we might restrict the values to the set {0,1,2,3} representing {*North, East, South, West*}. In either case an integer encoding is probably more suitable than a binary encoding. When designing the encoding and variation operators, it is worth considering whether there are any natural relations between the possible values that an attribute can take. This might be obvious for **ordinal attributes** such as integers (2 is more like 3 than it is 389), but for **cardinal attributes** such as the compass points above, there may not be a natural ordering.[3]

To give a well-known example of where there is no natural ordering, let us consider the graph k-colouring problem. Here we are given a set of points (vertices) and a list of connections between them (edges). The task is to assign one of k colours to each vertex, so that no two vertices which are connected by an edge share the same colour. For this problem there is no natural ordering: 'red' is no more like 'yellow' than 'blue', as long as they are different. In fact, we could assign the colours to the k integers representing them in any order, and still get valid equivalent solutions.

4.3.1 Mutation for Integer Representations

For integer encodings there are two principal forms of mutation used, both of which mutate each gene independently with user-defined probability p_m.

Random Resetting Here the bit-flipping mutation of binary encodings is extended to random resetting: in each position independently, with probability p_m, a new value is chosen at random from the set of permissible values. This is the most suitable operator to use when the genes encode for cardinal attributes, since all other gene values are equally likely to be chosen.

Creep Mutation This scheme was designed for ordinal attributes and works by adding a small (positive or negative) value to each gene with probability p. Usually these values are sampled randomly for each position, from a distribution that is symmetric about zero, and is more likely to generate small changes than large ones. It should be noted that creep mutation requires a number of parameters controlling the distribution from which the random numbers are drawn, and hence the size of the *steps* that mutation takes in the search space. Finding appropriate settings for these parameters may not be easy, and it is sometimes common to use more than one mutation operator in tandem from integer-based problems. For example, in [98] both a "big creep" and a "little creep" operator are used. Alternatively, random resetting might be used with low probability, in conjunction with a creep operator that tended to make small changes relative to the range of permissible values.

[3] There are various naming conventions used to distinguish these two types of attributes. These are discussed further in Chap. 7 and displayed in Table 7.1.

4.3.2 Recombination for Integer Representation

For representations where each gene has a finite number of possible allele values (such as integers) it is normal to use the same set of operators as for binary representations. On the one hand, these operators are valid: the offspring would not fall outside the given genotype space. On the other hand, these operators are also sufficient: it usually does not make sense to consider 'blending' allele values of this sort. For example, even if genes represent integer values, averaging an even and an odd integer yields a non-integral result.

4.4 Real-Valued or Floating-Point Representation

Often the most sensible way to represent a candidate solution to a problem is to have a string of real values. This occurs when the values that we want to represent as genes come from a continuous rather than a discrete distribution — for example, if they represent physical quantities such as the length, width, height, or weight of some component of a design that can be specified within a tolerance smaller than integer values. A good example would be the satellite dish holder boom described in Sect. 2.4, where the design is encoded as a series of angles and spar lengths. Another example might be if we wished to use an EA to evolve the weights on the connections beween the nodes in an artificial neural network. Of course, on a computer the precision of these real values is actually limited by the implementation, so we will refer to them as floating-point numbers. The genotype for a solution with k genes is now a vector $\langle x_1, \ldots, x_k \rangle$ with $x_i \in \mathbb{R}$.

4.4.1 Mutation for Real-Valued Representation

For floating-point representations, it is normal to ignore the discretisation imposed by hardware and consider the allele values as coming from a continuous rather than a discrete distribution, so the forms of mutation described above are no longer applicable. Instead it is common to change the allele value of each gene randomly within its domain given by a lower L_i and upper U_i bound,[4] resulting in the following transformation:

$$\langle x_1, \ldots, x_n \rangle \to \langle x'_1, \ldots, x'_n \rangle, \quad \text{where} \quad x_i, x'_i \in [L_i, U_i].$$

As with integer representations, two types can be distinguished according to the probability distribution from which the new gene values are drawn: uniform and nonuniform mutation.

[4] We assume here that the domain of each variable is a single interval $[L_i, U_i] \subseteq \mathbb{R}$. The generalisation to a union of disjoint intervals is straightforward.

Uniform Mutation For this operator the values of x'_i are drawn uniformly randomly from $[L_i, U_i]$. This is the most straightforward option, analogous to bit-flipping for binary encodings and the random resetting for integer encodings. It is normally used with a positionwise mutation probability.

Nonuniform Mutation Perhaps the most common form of nonuniform mutation used with floating-point representations takes a form analogous to the creep mutation for integers. It is designed so that usually, but not always, the amount of change introduced is small. This is achieved by adding to the current gene value an amount drawn randomly from a Gaussian distribution with mean zero and user-specified standard deviation, and then curtailing the resulting value to the range $[L_i, U_i]$ if necessary. This distribution, shown in Eq. 4.1, has the feature that the probability of drawing a random number with any given magnitude is a rapidly decreasing function of the standard deviation σ. Approximately two thirds of the samples drawn will lie within plus or minus one standard deviation, which means that most of the changes made will be small, but there is nonzero probability of generating very large changes since the tail of the distribution never reaches zero. Thus the σ value is a parameter of the algorithm that determines the extent to which given values x_i are perturbed by the mutation operator. For this reason σ is often called the **mutation step size**. It is normal practice to apply this operator with probability one per gene, and instead the mutation parameter is used to control the standard deviation of the Gaussian and hence the probability distribution of the step sizes taken.

$$p(\Delta x_i) = \frac{1}{\sigma\sqrt{2\pi}} \cdot e^{-\frac{(\Delta x_i - \xi)^2}{2\sigma^2}}. \tag{4.1}$$

An alternative to the Gaussian distribution is the use of a Cauchy distribution, which has a 'fatter' tail. That is, the probabilities of generating larger values are slightly higher than for a Gaussian with the same standard deviation [469].

4.4.2 Self-adaptive Mutation for Real-Valued Representation

As described above, non-uniform mutation applied to continuous variables is usually done by adding some random variables from a Gaussian distribution, with zero mean and a standard deviation which controls the mutation step size. The concept of **self-adaptation** represents a solution to the problem of how to adapt the step-sizes, which has been successfully demonstrated in many domains, not only for real-valued, but also for binary and integer search spaces [24]. The essential feature is that the step sizes are also included in the chromosomes and they themselves undergo variation and selection.

Details on how to mutate the value of σ are given below. The key concept is that the mutation step sizes are not set by the user; rather the σ coevolves with the solutions (the $\bar{x}$ part). In order to achieve this behaviour it is essential to modify the value of σ first, and then mutate the x_i values with the new σ value. The rationale behind this is that a new individual $\langle \bar{x}', \sigma' \rangle$ is effectively evaluated twice. Primarily, it is evaluated directly for its viability during survivor selection based on $f(\bar{x}')$. Second, it is evaluated for its ability to create good offspring. This happens indirectly: a given step size evaluates favourably if the offspring generated by using it prove viable (in the first sense). Thus, an individual $\langle \bar{x}', \sigma' \rangle$ represents both a good $\bar{x}'$ that survived selection and a good σ' that proved successful in generating this good $\bar{x}'$ from $\bar{x}$.

The alert reader may have noticed that there is an important underlying assumption behind the idea of using varying mutation step sizes. Namely, we assume that under different circumstances different step sizes will behave differently: some will be better than others. These circumstances can be given various interpretations. For instance, we might consider time and distinguish different stages within the evolutionary search process and expect that different mutation strategies would be appropriate in different stages. Self-adaptation can then be a mechanism adjusting the mutation strategy as the search is proceeding. Alternatively, we can consider space and observe that the local vicinity of an individual, i.e., the shape of the fitness landscape in its neighbourhood, determines what good mutations are: those that jump into the direction of fitness increase. Assigning a separate mutation strategy to each individual, which coevolves with it, opens the possibility to learn and use a mutation operator suited for the local topology. Issues related to these considerations are treated extensively in the chapter on parameter control, Chap. 8. In the following we describe three special cases of self-adaptive mutation in more detail.

Uncorrelated Mutation with One Step Size In the case of uncorrelated mutation with one step size, the same distribution is used to mutate each x_i, therefore we only have one strategy parameter σ in each individual. This σ is mutated each time step by multiplying it by a term e^{Γ}, with Γ a random variable drawn each time from a normal distribution with mean 0 and standard deviation τ. Since $N(0, \tau) = \tau \cdot N(0, 1)$, the mutation mechanism is thus specified by the following formulas:

$$\sigma' = \sigma \cdot e^{\tau \cdot N(0,1)}, \tag{4.2}$$

$$x_i' = x_i + \sigma' \cdot N_i(0, 1). \tag{4.3}$$

Furthermore, since standard deviations very close to zero are unwanted (they will have on average a negligible effect), the following boundary rule is used to force step sizes to be no smaller than a threshold:

$$\sigma' < \varepsilon_0 \Rightarrow \sigma' = \varepsilon_0.$$

In these formulas $N(0, 1)$ denotes a draw from the standard normal distribution, while $N_i(0, 1)$ denotes a separate draw from the standard normal distribution for each variable i. The proportionality constant τ is an external parameter to be set by the user. It is usually inversely proportional to the square root of the problem size:

$$\tau \propto 1/\sqrt{n}.$$

The parameter τ can be interpreted as a kind of **learning rate**, as in neural networks. Bäck [22] explains the reasons for mutating σ by multiplying with a variable with a lognormal distribution as follows:

- Smaller modifications should occur more often than large ones.
- Standard deviations have to be greater than 0.
- The median (0.5-quantile) should be 1, since we want to multiply the σ.
- Mutation should be neutral on average. This requires equal likelihood of drawing a certain value and its reciprocal value, for all values.

The lognormal distribution satisfies all these requirements.

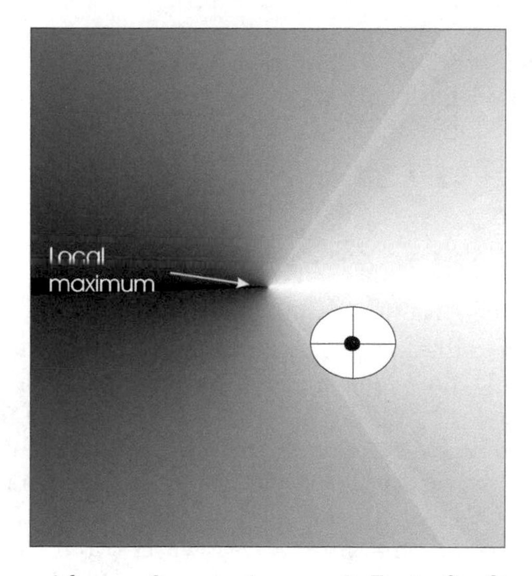

Fig. 4.4. Mutation with $n = 2, n_\sigma = 1, n_\alpha = 0$. Part of a fitness landscape with a *conical shape* is shown. The *black dot* indicates an individual. Points where the offspring can be placed with a given probability form a *circle*. The probability of moving along the y-axis (little effect on fitness) is the same as that of moving along the x-axis (large effect on fitness)

Figure 4.4 shows the effects of mutation in two dimensions. That is, we have an objective function $\mathbb{R}^2 \rightarrow \mathbb{R}$, and individuals are of the form $\langle x, y, \sigma \rangle$. Since there is only one σ, the mutation step size is the same in each direction and the points in the search space where the offspring can be placed with a given probability form a circle around the individual to be mutated.

Uncorrelated Mutation with n Step Sizes The motivation behind using n step sizes is the wish to treat dimensions differently. In particular, we want to be able to use different step sizes for different dimensions $i \in \{1, \ldots, n\}$. The reason for this is the trivial observation that the fitness landscape can have a different slope in one direction (along axis i) than in another direction (along axis j). The solution is straightforward: each basic chromosome $\langle x_1, \ldots, x_n \rangle$ is extended with n step sizes, one for each dimension, resulting in $\langle x_1, \ldots, x_n, \sigma_1, \ldots, \sigma_n \rangle$. The mutation mechanism is now specified as follows:

$$\sigma'_i = \sigma_i \cdot e^{\tau' \cdot N(0,1) + \tau \cdot N_i(0,1)}, \tag{4.4}$$
$$x'_i = x_i + \sigma'_i \cdot N_i(0,1), \tag{4.5}$$

where $\tau' \propto 1/\sqrt{2n}$, and $\tau \propto 1/\sqrt{2\sqrt{n}}$. Once again a boundary rule is applied to prevent standard deviations very close to zero.

$$\sigma'_i < \varepsilon_0 \Rightarrow \sigma'_i = \varepsilon_0.$$

Notice that the mutation formula for σ is different from that in Eq. (4.2). The present mutation mechanism is based on a finer granularity. Instead of the individual level (each individual $\bar{x}$ having its own σ) it works on the coordinate level (one σ_i for each x_i in $\bar{x}$). The corresponding straightforward modification of Eq. (4.2) is

$$\sigma'_i = \sigma_i \cdot e^{\tau \cdot N_i(0,1)},$$

but ES use Eq. (4.4). Technically, this is correct since the sum of two normally distributed variables is also normally distributed, hence the resulting distribution is still lognormal. The conceptual motivation is that the common base mutation $e^{\tau' \cdot N(0,1)}$ allows for an overall change of the mutability, guaranteeing the preservation of all degrees of freedom, while the coordinate-specific $e^{\tau \cdot N_i(0,1)}$ provides the flexibility to use different mutation strategies in different directions.

In Fig. 4.5 the effects of mutation are shown in two dimensions. Again, we have an objective function $\mathbb{R}^2 \to \mathbb{R}$, but the individuals now have the form $\langle x, y, \sigma_x, \sigma_y \rangle$. Since the mutation step sizes can differ in each direction (x and y), the points in the search space where the offspring can be placed with a given probability form an ellipse around the individual to be mutated. The axes of such an ellipse are parallel to the coordinate axes, with the length along axis i proportional to the value of σ_i.

Correlated Mutations The second version of mutation discussed above introduced different standard deviations for each axis, but this only allows ellipses orthogonal to the axes. The rationale behind correlated mutations is to allow the ellipses to have any orientation by rotating them with a rotation (covariance) matrix C.

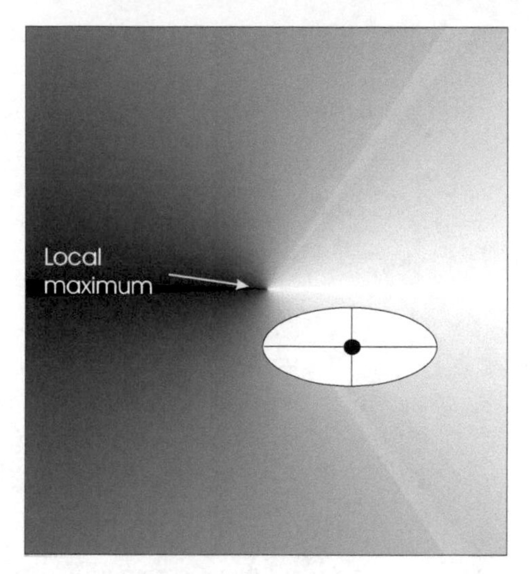

Fig. 4.5. Mutation with $n = 2, n_\sigma = 2, n_\alpha = 0$. Part of a fitness landscape with a *conical shape* is shown. The *black dot* indicates an individual. Points where the offspring can be placed with a given probability form an *ellipse*. The probability of moving along the x-axis (large effect on fitness) is larger than that of moving along the y-axis (little effect on fitness)

The probability density function for $\overline{\Delta x}$ replacing Eq. (4.1) now becomes

$$p(\overline{\Delta x}) = \frac{e^{-\frac{1}{2}\overline{\Delta x}^T \cdot C^{-1} \cdot \overline{\Delta x}}}{(\det C \cdot (2\pi)^n)^{1/2}},$$

with C the covariance matrix with entries

$$c_{ii} = \sigma_i^2, \tag{4.6}$$

$$c_{ij,i\neq j} = \begin{cases} 0 & \text{no correlations,} \\ \frac{1}{2}(\sigma_i^2 - \sigma_j^2)\tan(2\alpha_{ij}) & \text{correlations.} \end{cases} \tag{4.7}$$

The relation between covariance and rotation angle is as follows:

$$\tan(2\alpha_{ij}) = \frac{2c_{ij}}{\sigma_i^2 - \sigma_j^2},$$

which explains Eq. (4.7). This formula is derived from the trigonometric properties of rotations. A rotation in two dimensions is a multiplication with the matrix

$$\begin{pmatrix} \cos(\alpha_{ij}) & -\sin(\alpha_{ij}) \\ \sin(\alpha_{ij}) & \cos(\alpha_{ij}) \end{pmatrix}.$$

A rotation in more dimensions can be performed by a successive series of 2D rotations, i.e., matrix multiplications.

The complete mutation mechanism is described by the following equations:

$$\sigma'_i = \sigma_i \cdot e^{\tau' \cdot N(0,1) + \tau \cdot N_i(0,1)},$$
$$\alpha'_j = \alpha_j + \beta \cdot N_j(0,1),$$
$$\overline{x}' = \overline{x} + \overline{N}(\overline{0}, C'),$$

where $n_\alpha = \frac{n \cdot (n-1)}{2}$, $j \in 1, \ldots, n_\alpha$. The other constants are usually taken as: $\tau \propto 1/\sqrt{2\sqrt{n}}$, $\tau' \propto 1/\sqrt{2n}$, and $\beta \approx 5^o$.

The object variables $\overline{x}$ are now mutated by adding $\overline{\Delta x}$ drawn from an n-dimensional normal distribution with covariance matrix C'. The C' in the formula is the old C after mutation of the α values (and recalculation of covariances). The σ_i are mutated in the same way as before: with a multiplication by a log-normal variable, which consists of a global and an individual part. The α_j are mutated with an additive, normally distributed variation, similar to mutation of object variables.

We also have a boundary rule for the α_j values. The rotation angles should lie in the range $[-\pi, \pi]$, so the new value is simply mapped circularly into the feasible range:

$$|\alpha'_j| > \pi \Rightarrow \alpha'_j = \alpha'_j - 2\pi \ \mathrm{sign}(\alpha'_j).$$

Fig. 4.6 shows the effects of correlated mutations in two dimensions. The individuals now have the form $\langle x, y, \sigma_x, \sigma_y, \alpha_{x,y} \rangle$, and the points in the search space where the offspring can be placed with a given probability form a rotated ellipse around the individual to be mutated, where again the axis lengths are proportional to the σ values.

Table 4.1 summarises three possible common settings for self-adaptive mutation regarding the length and structure of the individuals. Simply considering the size of the representation of the individuals in each scheme, i.e., the number of values that need to be learned by the algorithm as it evolves (let alone their complex interrelationships) brings home an important point: we can get nothing for free! In other words, what we must consider is that as the ability of the algorithm to adapt the nature of its search according to the local topology increases, so too does the scale of the learning task. To simplify matters a little, as we increase the precision with which we can specify the shape of the lines of equiprobable mutations, so we increase the number of different options which should be tried. Since the merits of these different possibilities are evaluated indirectly, i.e., by applying them and gauging the relative fitness of the individuals created, it is reasonable to conclude that an increased number of function evaluations will be needed to learn good search strategies as the complexity of the mutation operator increases.

While this may sound a little pessimistic, it is also worth noting that it is easy to imagine a situation where the extra complexity is required, for example, if the landscape contains a 'ridge' of increasing fitness, perhaps running at

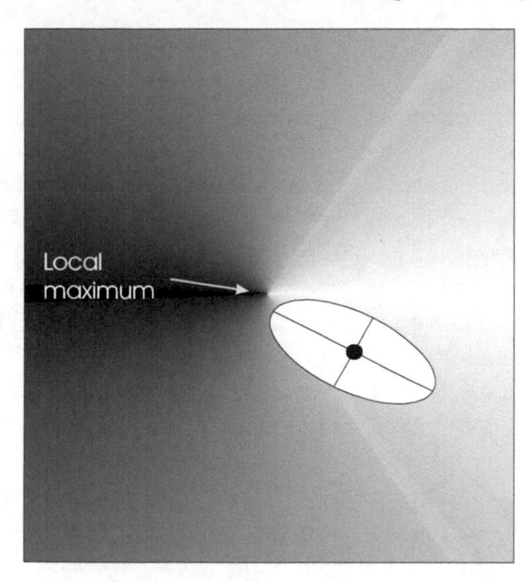

Fig. 4.6. Correlated mutation: $n = 2, n_\sigma = 2, n_\alpha = 1$. Part of a fitness landscape with a *conical shape* is shown. The *black dot* indicates an individual. Points where the offspring can be placed with a given probability form a *rotated ellipse*. The probability of generating a move in the direction of the steepest ascent (largest effect on fitness) is now larger than that for other directions

an angle to the co-ordinate axis. In short, there are no fixed recommendations about which scheme to use, but a common approach is to start with uncorrelated mutation with n σ values and then try moving to a simpler model if good results are obtained but too slowly (or if the σ_i all evolve to similar values), or to the more complex model if the results are not of good enough quality.

n_σ	n_α	Structure of individuals	Remark
1	0	$\langle x_1, \ldots, x_n, \sigma \rangle$	Standard mutation
n	0	$\langle x_1, \ldots, x_n, \sigma_1, \ldots, \sigma_n \rangle$	Standard mutations
n	$n \cdot (n-1)/2$	$\langle x_1, \ldots, x_n, \sigma_1, \ldots, \sigma_n, \alpha_1, \ldots, \alpha_{n \cdot (n-1)/2} \rangle$	Correlated mutations

Table 4.1. Some possible settings of n_σ and n_α for different mutation operators

Self-adaptive mutation mechanisms have been used and studied for decades in EC. Besides experimental evidence, showing that an EA with self-adaptation outperforms the same algorithm without self-adaptation, there are also theoretical results showing that self-adaptation works [52]. Theoretical and experimental results can neatly complement each other in this area if experimentally obtained mutation step sizes show a good match with the theoretically derived optimal values. Unfortunately, for a complex problem and/or algorithm a theoretical analysis is infeasible. However, for simple objective functions theoretically optimal mutation step sizes can be calculated (in light of some performance criterion, e.g., progress rate during a run) and compared to step sizes obtained during a run of the EA in question.

Theoretical and experimental results agree on the fact that for a successful run the σ values must decrease over time. The intuitive explanation for this is that in the beginning of a search process a large part of the search space has to be sampled in an explorative fashion to locate promising regions (with good fitness values). Therefore, large mutations are appropriate in this phase. As the search proceeds and optimal values are approached, only fine tuning of the given individuals is needed; thus smaller mutations are required.

Another kind of convincing evidence for the power of self-adaptation is provided in the context of changing fitness landscapes. In this case, where the objective function is changing, the evolutionary process is aiming at a moving target. When the objective function changes, the given individuals may have a low fitness, since they have been adapted to the old objective function. Thus, the present population needs to be reevaluated, and the search space re-explored. Often the mutation step sizes will prove ill-adapted: they are too low for the new exploration phase required. The experiment presented in [217] illustrates how self-adaptation is able to reset the step sizes after each change in the objective function (Fig. 4.7).

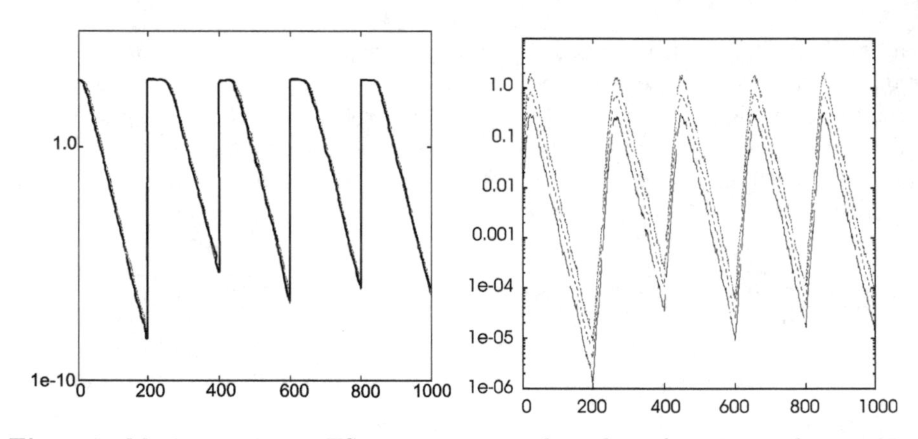

Fig. 4.7. Moving optimum ES experiment on the sphere function with $n = 30$, $n_\sigma = 1$. The location of the optimum is changed after every 200 generations (x-axes) with a clear effect on the average best objective function values (y-axis, left) in the given population. Self-adaptation is adjusting the step sizes (y-axis, right) with a small delay to larger values appropriate for exploring the new fitness landscape, whereafter the values of σ start decreasing again as the population approaches the new optimum

Over recent decades much experience has been gained over self-adaptation in Evolutionary Algorithms, in particular in Evolution Strategies. The accumulated knowledge has identified necessary conditions for self-adaptation:

1. $\mu > 1$ so that different strategies are present
2. generation of an offspring surplus: $\lambda > \mu$
3. a not too strong selective pressure (heuristic: $\lambda/\mu = 7$, e.g., (15,100))
4. (μ, λ)-selection (to guarantee extinction of misadapted individuals)
5. recombination, usually intermediate, of strategy parameters

4.4.3 Recombination Operators for Real-Valued Representation

In general, we have three options for recombining two floating-point strings. First, using an analogous operator to those used for bit-strings, but now split between floats. In other words, an allele is one floating-point value instead of one bit. This has the disadvantage (shared with all of the recombination operators described above) that only mutation can insert new values into the population, since recombination only gives us new combinations of existing values. Recombination operators of this type for floating-point representations are known as **discrete recombination** and have the property that if we are creating an offspring z from parents x and y, then the allele value for gene i is given by $z_i = x_i$ or y_i with equal likelihood.

Second, using an operator that, in each gene position, creates a new allele value in the offspring that lies between those of the parents. Using the terminology above, we have $z_i = \alpha x_i + (1 - \alpha) y_i$ for some α in [0,1]. In this way, recombination is now able to create new gene material, but it has the disadvantage that as a result of the averaging process the range of the allele values in the population for each gene is reduced. Operators of this type are known as **intermediate** or **arithmetic recombination**.

Third, using an operator that in each position creates a new allele value in the offspring which is close to that of one of the parents, but may lie outside them (i.e., bigger than the larger of the two values, or smaller than the lesser). Operators of this type can create new material without restricting the range. Operators of this type are known as **blend recombination**.

Three types of arithmetic recombination are described in [295]. In all of these, the choice of the parameter α is sometimes made at random over [0,1], but in practice it is common to use a constant value, often 0.5 (in which case we have **uniform arithmetic recombination**).

Simple Arithmetic Recombination First pick a recombination point k. Then, for child 1, take the first k floats of parent 1 and put them into the child. The rest is the arithmetic average of parent 1 and 2:

$$\text{Child 1: } \langle x_1, \ldots, x_k, \alpha \cdot y_{k+1} + (1 - \alpha) \cdot x_{k+1}, \ldots, \alpha \cdot y_n + (1 - \alpha) \cdot x_n \rangle.$$

Child 2 is analogous, with x and y reversed (Fig. 4.8, top).

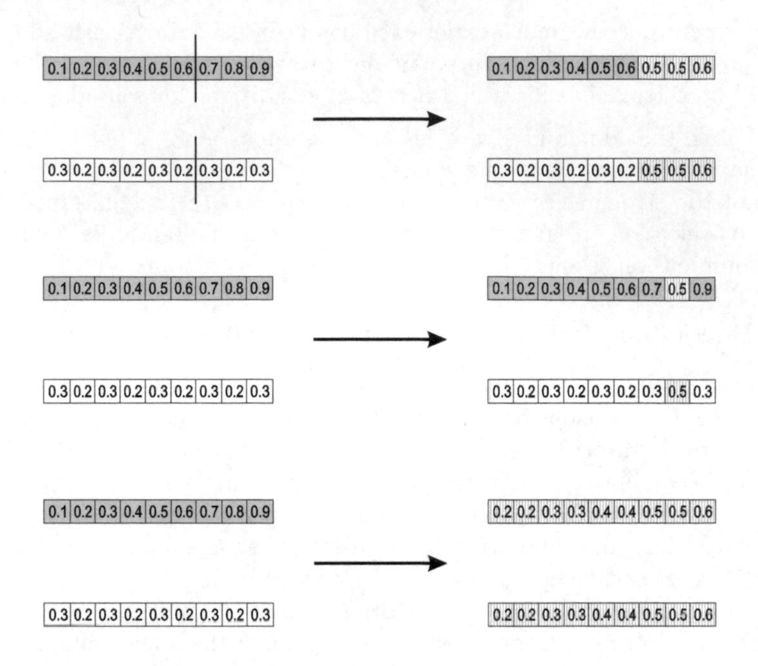

Fig. 4.8. Simple arithmetic recombination with $k = 6, \alpha = 1/2$ (top), single arithmetic recombination with $k = 8, \alpha = 1/2$ (middle), whole arithmetic recombination with $\alpha = 1/2$ (bottom).

Single Arithmetic Recombination Pick a random allele k. At that position, take the arithmetic average of the two parents. The other points are the points from the parents, i.e.:

$$\text{Child 1: } \langle x_1, \ldots, x_{k-1}, \alpha \cdot y_k + (1-\alpha) \cdot x_k, x_{k+1}, \ldots, x_n \rangle.$$

The second child is created in the same way with x and y reversed (Fig. 4.8, middle).

Whole Arithmetic Recombination This is the most commonly used operator and works by taking the weighted sum of the two parental alleles for each gene, i.e.:

$$\text{Child 1} = \alpha \cdot \bar{x} + (1-\alpha) \cdot \bar{y}, \qquad \text{Child 2} = \alpha \cdot \bar{y} + (1-\alpha) \cdot \bar{x}.$$

This is illustrated in Fig. 4.8, bottom. As the example shows, if $\alpha = 1/2$ the two offspring will be identical for this operator.

Blend Crossover Blend Crossover ($BLX - \alpha$) was introduced in [160] as a way of creating offspring in a region that is bigger than the (n-dimensional) rectangle spanned by the parents. The extra space is proportional to the

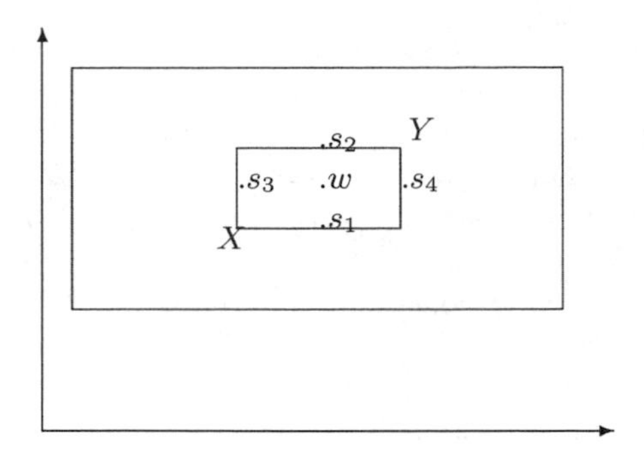

Fig. 4.9. Possible offspring from different recombination operators for two real-valued parents X and Y. $\{s_1, \ldots, s_4\}$ are the four possible offspring from single arithmetic recombination with $\alpha = 0.5$. w is the offspring from whole arithmetic recombination with $\alpha = 0.5$ and the *inner* box represents all the possible offspring positions as α is varied. The *outer dashed* box shows all possible offspring positions for blend crossover with $\alpha = 0.5$ ($BLX - 0.5$), each position being equally likely.

distance between the parents and it varies per coordinate. If we have two parents x and y and assume that in position i the value $x_i < y_i$ then the difference $d_i = y_i - x_i$ and the range for the ith value in the child z is $[x_i - \alpha \cdot d_i, x_i + \alpha \cdot d_i]$. To create a child we can sample a random number u uniformly from $[0, 1]$, calculate $\gamma = (1 - 2\alpha)u - \alpha$, and set:

$$z_i = (1 - \gamma)x_i + \gamma y_i$$

Interestingly, the original authors reported best results with $\alpha = 0.5$, where the chosen values are equally likely to lie inside the two parent values as outside, so balancing exploration and exploitation.

Figure 4.9 illustrates the difference between single arithmetic recombination, whole arithmetic combination and Blend Crossover, with in each case the value of α set to 0.5. More recent methods such as Simulated Binary Crossover [111, 113] have built on Blend Crossover, so that rather than selecting offspring values uniformly from a range around each parent values, they are selected from a distribution which is more likely to create small changes, and the distribution is controlled by the distance between the parents.

4.5 Permutation Representation

Many problems naturally take the form of deciding on the order in which a sequence of events should occur. While other forms do occur (for example, decoder functions based on unrestricted integer representations [28, 201] or "floating keys" based on real-valued representations [27, 44]), the most natural representation of such problems is as a permutation of a fixed set of values that can be represented as integers. One immediate consequence is that while a binary, or simple integer, representation allows numbers to occur more than once, such sequences of integers will not represent valid permutations. It is clear therefore that when choosing or designing variation operators to work with solutions that are represented as permutations, we require them to preserve the permutation property that each possible allele value occurs exactly once in the solution. We previously described one example, when we designed an EA for solving the N-queens problem efficiently, by representing each solution as a list of the rows on which each queen was positioned (with each on a different column), and insisted that these be a permutation so that no two queens shared the same row.

When choosing variation operators it is worth bearing in mind that there are actually two classes of problems that are represented by permutations. In the first of these, the *order* in which events occur is important. This might happen when the events use limited resources or time, and a typical example of this sort of problem is the production scheduling problem. This is the common problem of deciding in which order a series of times should be manufactured on a set of machines, where there may be dependencies between products, for example, there might be different set-up times between products, or one might be a component of another. As an example, it might be better for widget 1 to be produced before widgets 2 and 3, which in turn might be preferably produced before widget 4, no matter how far in advance this is done. In this case it might well be that the sequences [1,2,3,4] and [1,3,2,4] have similar fitness, and are much better than, for example, [4,3,2,1].

Another type of problem depends on *adjacency*, and is typified by the travelling salesperson problem (TSP). The problem is to find a complete tour of n given cities of minimal length. The search space for this problem is huge: there are $(n\text{-}1)!$ different routes possible for n given cities (for the asymmetric case counting back and forth as two routes).[5] For $n = 30$ there are approximately 10^{32} different tours. Labelling the cities $1, 2, \dots, n$, a complete tour is a permutation, so that for $n = 4$, the routes [1,2,3,4] and [3,4,2,1] are both valid. The vital point here is that it is the links between cities that are important. The difference from order-based problems can clearly be seen if we consider that the starting point of the tour is also not important, thus [1,2,3,4], [2,3,4,1], [3,4,1,2], and [4,1,2,3] are all equivalent. Many examples of this class are also symmetric, so that [4,3,2,1] and so on are also equivalent.

Finally, we should mention that there are two possible ways to encode a permutation. In the first (most commonly used) of these the ith element of the representation denotes the event that happens in that place in the sequence (or the ith destination visited). In the second, the value of the ith element denotes the position in the sequence in which the ith event happens. Thus for the four cities [A,B,C,D], and the permutation [3,1,2,4], the first encoding denotes the tour [C,A,B,D] and the second [B,C,A,D].

---

[5] These comments about problem size apply to all permutation problems.

4.5.1 Mutation for Permutation Representation

For permutation representations, it is no longer possible to consider each gene independently, rather finding legal mutations is a matter of moving alleles around in the genome. This has the immediate consequence that the mutation parameter is interpreted as the probability that the *chromosome* undergoes mutation, rather than that a single gene in the chromosome is altered. The three most common forms of mutation used for order-based problems were first described in [423]. Whereas the first three operators below (in particular insertion) work by making small changes to the order in which allele values occur, for adjacency-based problems these can cause huge numbers of links to be broken, and so inversion is more commonly used.

Swap Mutation Two positions (genes) in the chromosome are selected at random and their allele values swapped. This is illustrated in Fig. 4.10 (top), where the values in positions two and five have been swapped.
Insert Mutation Two alleles are selected at random and the second moved next to the first, shuffling along the others to make room. This is illustrated in Fig. 4.10 (middle), where the values two and five have been chosen.
Scramble Mutation Here the entire chromosome, or some randomly chosen subset of values within it, have their positions scrambled. This is illustrated in Fig. 4.10 (bottom), where the values from two to five have been chosen.

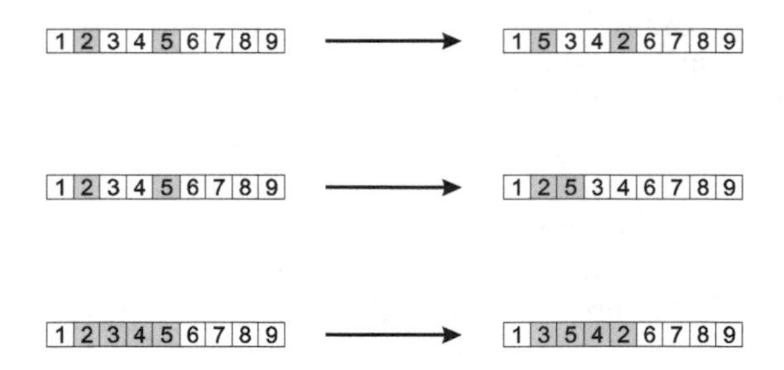

Fig. 4.10. Swap (top), insert (middle), and scramble mutation (bottom).

Inversion Mutation Inversion mutation works by randomly selecting two positions in the chromosome and reversing the order in which the values appear between those positions. It effectively breaks the chromosome into three parts, with all links inside a part being preserved, and only the two links between the parts being broken. The inversion of a randomly chosen substring is the thus smallest change that can be made to an adjacency-based problem, and all other changes can be easily constructed as a series of inversions. The ordering of the search space induced by this operator thus forms a natural basis for considering this class of problems, equivalent to the Hamming space for binary problem representations. It is the basic move behind the 2-opt search heuristic for TSP [271], and by extension k-opt. This operator is illustrated in Fig. 4.11, where the substring between positions two and five was inverted.

1	2	3	4	5	6	7	8	9

→

1	5	4	3	2	6	7	8	9

Fig. 4.11. Inversion mutation

4.5.2 Recombination for Permutation Representation

At first sight, permutation-based representations present particular difficulties for the design of recombination operators, since it is not generally possible simply to exchange substrings between parents and still maintain the permutation property. However, this situation is alleviated when we consider what it is that the solutions actually represent, i.e., either an order in which elements occur, or a set of moves linking pairs of elements. A number of specialised recombination operators have been designed for permutations, which aim at transmitting as much as possible of the information contained in the parents, especially that held in common. We shall concentrate here on describing two of the best known and most commonly used operators for each subclass of permutation problems.

Partially Mapped Crossover (PMX) was first proposed by Goldberg and Lingle as a recombination operator for the TSP in [192], and has become one of the most widely used operators for adjacency-type problems. Over the years many slight variations of PMX appeared in the literature; here we use Whitley's definition from [452], which works as follows (Figs. 4.12–4.14).

1. Choose two crossover points at random, and copy the segment between them from the first parent (P1) into the first offspring.
2. Starting from the first crossover point look for elements in that segment of the second parent (P2) that have not been copied.
3. For each of these (say i), look in the offspring to see what element (say j) has been copied in its place from P1.
4. Place i into the position occupied by j in P2, since we know that we will not be putting j there (as we already have it in our string).
5. If the place occupied by j in P2 has already been filled in the offspring by an element k, put i in the position occupied by k in P2.

6. Having dealt with the elements from the crossover segment, the remaining positions in this offspring can be filled from P2, and the second child is created analogously with the parental roles reversed.

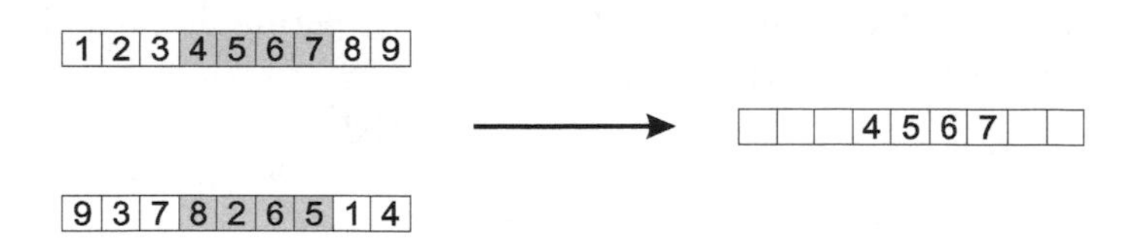

Fig. 4.12. PMX, step 1: copy randomly selected segment from first parent into offspring

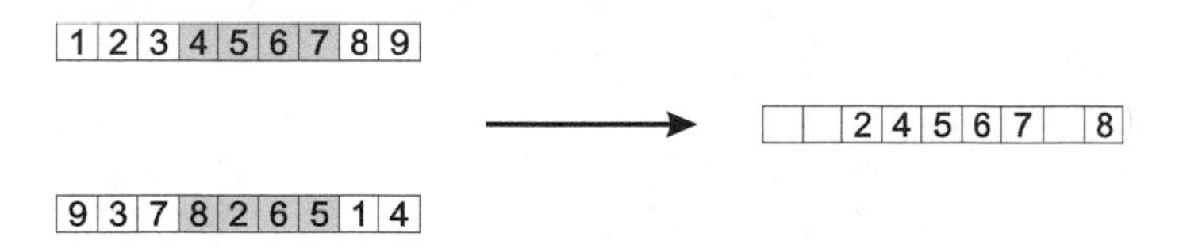

Fig. 4.13. PMX, step 2: consider in turn the placement of the elements that occur in the middle segment of parent 2 but not parent 1. The position that 8 takes in P2 is occupied by 4 in the offspring, so we can put the 8 into the position vacated by the 4 in P2. The position of the 2 in P2 is occupied by the 5 in the offspring, so we look first to the place occupied by the 5 in P2, which is position 7. This is already occupied by the value 7, so we look to where this occurs in P2 and finally find a slot in the offspring that is vacant – the third. Finally, note that the values 6 and 5 occur in the middle segments of both parents.

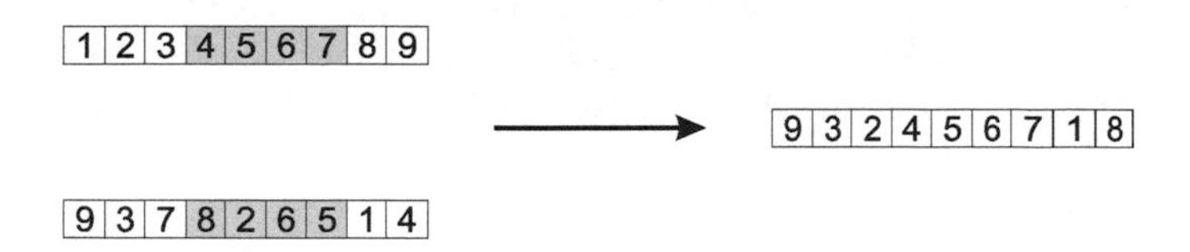

Fig. 4.14. PMX, step 3: copy remaining elements from second parent into same positions in offspring

Inspection of the offspring created shows that in this case six of the nine links present in the offspring are present in one or more of the parents. However, of the two edges {5–6} and {7–8} common to both parents, only the first is present in the offspring. Radcliffe [350] suggests that a desirable property of any recombination operator is that of *respect*, i.e., that any information carried in both parents should also be present in the offspring. A moment's reflection tells us that this is clearly true for all of the recombination operators described above for binary and integer representations, and for discrete recombination for floating-point representations, but as the example above shows, is not necessarily true of PMX. With this issue in mind, several other operators have been designed for adjacency-based permutation problems, of which the best known is described next.

Edge crossover is based on the idea that offspring should be created as far as possible using only edges that are present in (one of) the parents. It has undergone a number of revisions over the years. Here we describe the most commonly used version: edge-3 crossover after Whitley [452], which is designed to ensure that common edges are preserved.

In order to achieve this, an edge table (also known as an adjacency list) is constructed, which for each element lists the other elements that are linked to it in the two parents. A '+' in the table indicates that the edge is present in both parents. The operator works as follows:

1. Construct the edge table
2. Pick an initial element at random and put it in the offspring
3. Set the variable *current_element* = *entry*
4. Remove all references to *current_element* from the table
5. Examine list for *current_element*
   - If there is a common edge, pick that to be the next element
   - Otherwise pick the entry in the list which itself has the shortest list
   - Ties are split at random
6. In the case of reaching an empty list, the other end of the offspring is examined for extension; otherwise a new element is chosen at random

Clearly only in the last case will so-called foreign edges be introduced.

Edge-3 recombination is illustrated by the following example where the parents are the same two permutations used in the PMX example [1 2 3 4 5 6 7 8 9] and [9 3 7 8 2 6 5 1 4], giving the edge table seen in Table 4.2 and the construction illustrated in Table 4.3. Note that only one child per recombination is created by this operator.

Element	Edges	Element	Edges
1	2,5,4,9	6	2,5+,7
2	1,3,6,8	7	3,6,8+
3	2,4,7,9	8	2,7+, 9
4	1,3,5,9	9	1,3,4,8
5	1,4,6+		

Table 4.2. Edge crossover: example edge table

Choices	Element selected	Reason	Partial result
All	1	Random	[1]
2,5,4,9	5	Shortest list	[1 5]
4,6	6	Common edge	[1 5 6]
2,7	2	Random choice (both have two items in list)	[1 5 6 2]
3,8	8	Shortest list	[1 5 6 2 8]
7,9	7	Common edge	[1 5 6 2 8 7]
3	3	Only item in list	[1 5 6 2 8 7 3]
4,9	9	Random choice	[1 5 6 2 8 7 3 9]
4	4	Last element	[1 5 6 2 8 7 3 9 4]

Table 4.3. Edge crossover: example of permutation construction

Order crossover This operator was designed by Davis for order-based permutation problems [98]. It begins in a similar fashion to PMX, by copying a randomly chosen segment of the first parent into the offspring. However, it proceeds differently because the intention is to transmit information about *relative order* from the second parent.

1. Choose two crossover points at random, and copy the segment between them from the first parent (P1) into the first offspring.
2. Starting from the second crossover point in the second parent, copy the remaining unused numbers into the first child in the order that they appear in the second parent, wrapping around at the end of the list.
3. Create the second offspring in an analogous manner, with the parent roles reversed.

This is illustrated in Figs. 4.15 and 4.16.

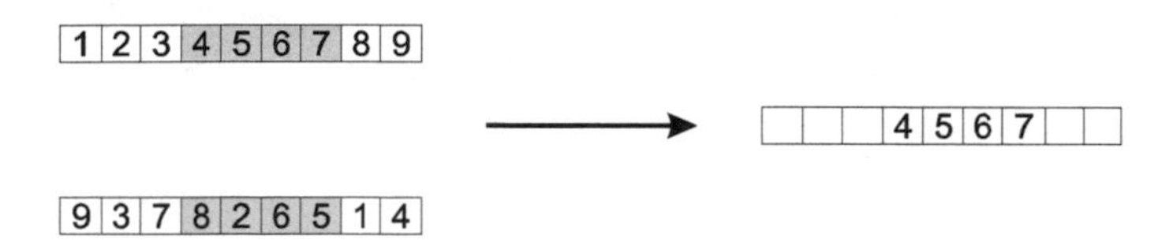

Fig. 4.15. Order crossover, step 1: copy randomly selected segment from first parent into offspring

Cycle Crossover The final operator that we will consider in this section is cycle crossover [325], which is concerned with preserving as much information as possible about the *absolute* position in which elements occur. The operator

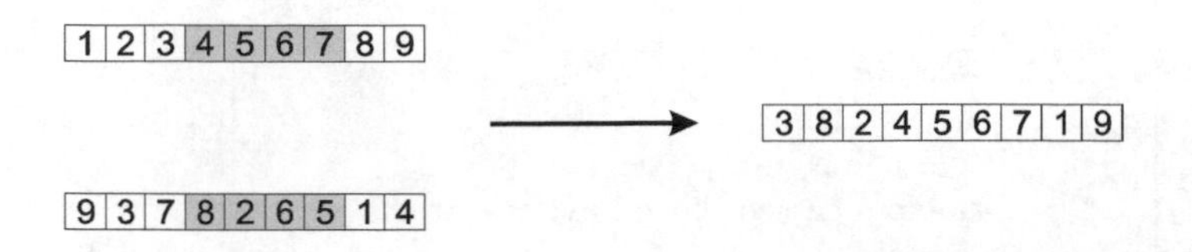

Fig. 4.16. Order crossover, step 2: copy rest of alleles in order they appear in second parent, treating string as toroidal

works by dividing the elements into *cycles*. A cycle is a subset of elements that has the property that each element always occurs paired with another element of the same cycle when the two parents are aligned. Having divided the permutation into cycles, the offspring are created by selecting alternate cycles from each parent. The procedure for constructing cycles is as follows:

1. Start with the first unused position and allele of P1
2. Look at the allele in the *same position* in P2
3. Go to the position with the *same allele* in P1
4. Add this allele to the cycle
5. Repeat steps 2 through 4 until you arrive at the first allele of P1

The complete operation of the operator is illustrated in Fig. 4.17.

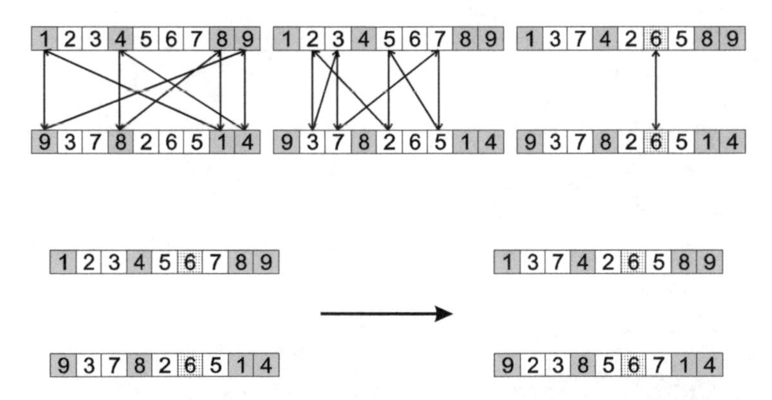

Fig. 4.17. Cycle crossover. Top: step 1- identification of cycles. Bottom: step 2- construction of offspring

4.6 Tree Representation

Trees are among the most general structures for representing objects in computing, and form the basis for the branch of evolutionary algorithms known as genetic programming (GP). In general, (parse) trees capture expressions in a given formal syntax. Depending on the problem at hand, and the users' perceptions on what the solutions must look like, this can be the syntax of arithmetic expressions, formulas in first-order predicate logic, or code written in a programming language. To illustrate the matter, let us consider one of each of these types of expressions.

- an arithmetic formula:

$$2 \cdot \pi + ((x+3) - \frac{y}{5+1}), \tag{4.8}$$

- a logical formula:

$$(x \wedge true) \rightarrow ((x \vee y) \vee (z \leftrightarrow (x \wedge y))), \tag{4.9}$$

- the following program:

```
i = 1;
while (i < 20)
{
        i = i+1;
}
```

Figures. 4.18 and 4.19 show the parse trees belonging to these expressions. These examples illustrate generally how parse trees can be used and interpreted.

Technically speaking, the specification of how to represent individuals boils down to defining the syntax of the trees, or equivalently the syntax of the symbolic expressions (**s-expressions**) they represent. This is commonly done by defining a **function set** and a **terminal set**. Elements of the terminal set are allowed as leaves, while symbols from the function set are internal nodes. For example, a suitable function and terminal set that allow the expression in Eq. (4.8) as syntactically correct is given in Table 4.4.

Function set	$\{+, -, \cdot, /\}$
Terminal set	$\mathbb{R} \cup \{x, y\}$

Table 4.4. Function and terminal set that allow the expression in Eq. (4.8) as syntactically correct

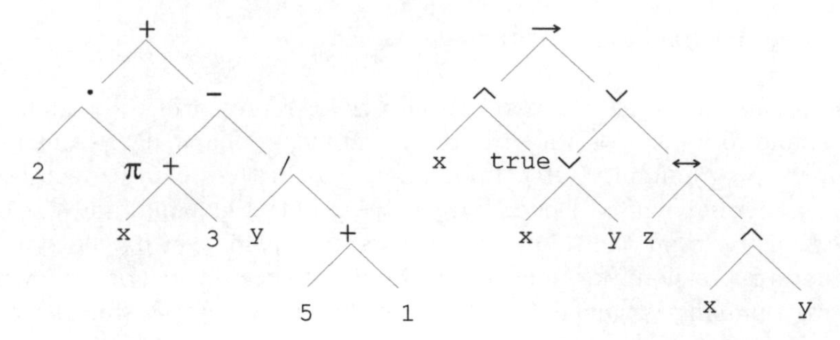

Fig. 4.18. Parse trees belonging to Eqs. (4.8) (left) and (4.9) (right)

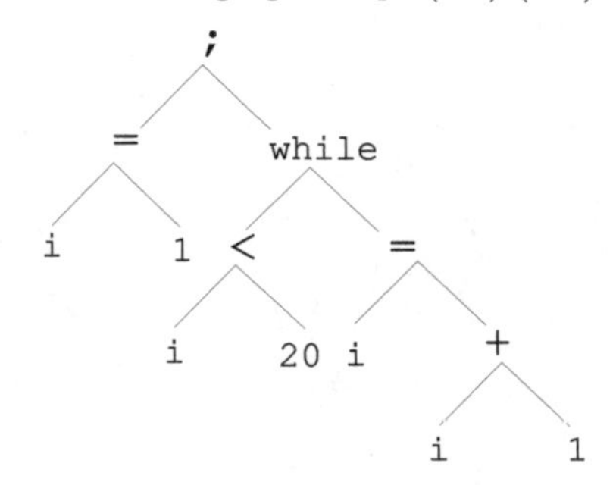

Fig. 4.19. Parse tree belonging to the above program

Strictly speaking, we should specify the arity (the number of attributes it takes) for each function symbol in the function set, but for standard arithmetic or logical functions this is often omitted. Similarly, a definition of correct expressions (trees) based on the function and terminal set should be given. However, as this follows the general way of defining terms in formal languages it is also often omitted. For the sake of completeness we provide it below:

- All elements of the terminal set T are correct expressions.
- If $f \in F$ is a function symbol with arity n and $e_1, \ldots, e_n$ are correct expressions, then so is $f(e_1, \ldots, e_n)$.
- There are no other forms of correct expressions.

Note that in this definition we do not distinguish different types of expressions; each function symbol can take any expression as argument. This feature is known as the **closure property**.

In practice, function symbols and terminal symbols are often typed and impose extra syntactic requirements. For instance, one might need both arithmetic and logical function symbols, e.g., to allow $(N = 2) \wedge (S > 80.000))$ as a correct expression. In this case it is necessary to enforce that an arithmetic (logical) function symbol only has arithmetic (logical) arguments, e.g., to exclude $N \wedge 80.000$ as a correct expression. This issue is addressed in strongly typed genetic programming [304].

4.6.1 Mutation for Tree Representation

The most common implementation of **tree-based mutation** works by selecting a node at random from the tree, and replacing the subtree starting there with a randomly generated tree. This newly created subtree is usually generated the same way as in the initial population, (Sect. 6.4), and so is subject to conditions on maximum depth and width. Figure 4.20 illustrates how the parse tree belonging to Eq. (4.8) (left) is mutated into one standing for $2 \cdot \pi + ((x + 3) - y)$. Note that since a node is selected at random to be the replacement point, and that as one goes down through a tree there are potentially more nodes at any given depth, the size (tree depth) of the child can exceed that of the parent tree.

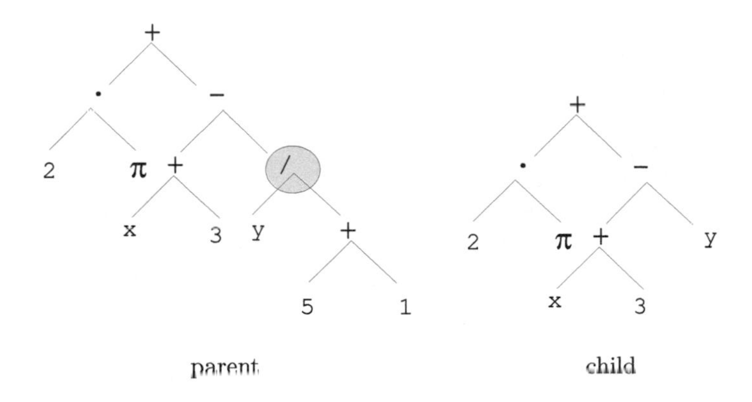

Fig. 4.20. Tree-based mutation illustrated: the node designated by a *circle* in the tree *on the left* is selected for mutation. The subtree staring at that node is replaced by a randomly generated tree, which is a leaf here

Tree-based mutation has two parameters:

- the probability of choosing mutation at the junction with recombination
- the probability of choosing an internal point within the parent as the root of the subtree to be replaced

It is remarkable that Koza's classic book on GP from 1992 [252] advises users to set the mutation rate at 0, i.e., it suggests that GP works *without* mutation. More recently Banzhaf et al. recommended 5% [37]. In giving mutation such a limited role, GP differs from other EA streams. The reason for this is the generally shared view that crossover has a large shuffling effect, acting in some sense as a macromutation operator [9]. The current GP practice uses low, but positive, mutation frequencies, even though some studies indicate that the common wisdom favouring an (almost) pure crossover approach might be misleading [275].

4.6.2 Recombination for Tree Representation

Tree-based recombination creates offspring by swapping genetic material among the selected parents. In technical terms, it is a binary operator creating two child trees from two parent trees. The most common implementation is **subtree crossover**, which works by interchanging the subtrees starting at two randomly selected nodes in the given parents. This is illustrated in Fig. 4.21. Note that the size (tree depth) of the children can exceed that of the parent trees. In this, recombination within GP differs from recombination in other EC dialects. Tree-based recombination has two parameters:

- the probability of choosing recombination at the junction with mutation
- the probability of choosing internal nodes as crossover points

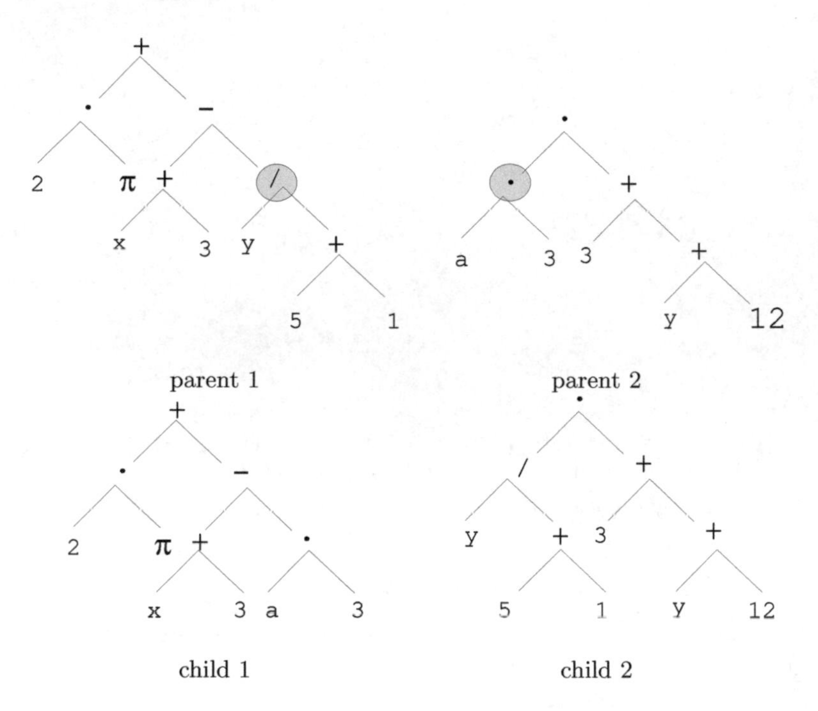

Fig. 4.21. Tree-based crossover illustrated: the nodes designated by a *circle* in the parent trees are selected to serve as crossover points. The subtrees staring at those nodes are swapped, resulting in two new trees, which are the children

For exercises and recommended reading for this chapter, please visit `www.evolutionarycomputation.org`.

5

Fitness, Selection, and Population Management

As explained in Chap. 3, there are two fundamental forces that form the basis of evolutionary systems: variation and selection. In this chapter we discuss the EA components behind the second one. Having discussed some typical population management models, and selection operators, we then go on to explicitly look at some situations where diversity is needed, such as multimodal problems, and some approaches to population management, and altering selection, that have been proposed to increase useful diversity.

5.1 Population Management Models

In the previous chapter we have focused on the way that potential solutions are represented to give a population of diverse individuals, and on the way that variation (recombination and mutation) operators work on those individuals to yield offspring. These offspring will generally inherit some of their parents' properties but also differ slightly from them, providing new potential solutions to be evaluated. We now turn our attention to the second important element of the evolutionary process – the differential survival of individuals to compete for resources and take part in reproduction, based on their relative fitness.

Two different models of population management are found in the literature: the **generational model** and the **steady-state model**. The generational model is the one used in the example in Sect. 3.3. In each generation we begin with a population of size μ, from which a **mating pool** of parents is selected. Every member of the pool is a copy of something in the population, but the proportions will probably differ, with (usually) more copies of the 'better' parents. Next, λ offspring are created from the mating pool by the application of variation operators, and evaluated. After each generation, the whole population is replaced by μ individuals selected from its offspring, which is called the next generation. In the model typically used within the Simple Genetic Algorithm, the population, mating pool and offspring are all the same size, so that each generation is replaced by all of its offspring. This restriction is not necessary: for example in the (μ, λ) Evolution Strategy, an excess of offspring is created (typically λ/μ is in the range 5–7) from which the next generation is selected on the basis of fitness.

In the steady-state model, the entire population is not changed at once, but rather a part of it. In this case, λ ($< \mu$) old individuals are replaced by λ new ones, the offspring. The proportion of the population that is replaced is called the **generational gap**, and is equal to λ/μ. Since its introduction in Whitley's GENITOR algorithm [460], the steady-state model has been widely studied and applied [105, 354, 442], often with $\lambda = 1$.

At this stage it is worth reiterating that the operators that are responsible for this competitive element of population management work on the basis of an individual's fitness. As a direct consequence, these selection and replacement operators work *independently* of the problem representation chosen. As was seen in the general description of an evolutionary algorithm at the start of Chap. 3, there are two points in the evolutionary cycle at which fitness-based competition can occur: during selection to take part in mating, and during the selection of individuals to survive into the next generation. We begin by describing the most commonly used methods for parent selection, but note that many of these can also be applied during the survival selection phase. As a final preliminary, please note that we will adopt a convention that we are trying to maximise fitness, and that fitness values are not negative. Often problems are expressed in terms of an objective function to be minimised, and sometimes negative fitness values occur. However, in all cases these can be mapped into the desired form by using an appropriate transformation.

5.2 Parent Selection

5.2.1 Fitness Proportional Selection

The principles of **fitness proportional selection** (FPS) were described in the simple example in Sect. 3.3. Recall that for each choice, the probability that an individual i is selected for mating depends on its *absolute* fitness value compared to the *absolute* fitness values of the rest of the population. Observing that the sum of the probabilities over the whole population must equal 1 the selection probability of individual i using FPS is $P_{FPS}(i) = f_i / \sum_{j=1}^{\mu} f_j$.

This selection mechanism was introduced in [220] and has been the topic of intensive study ever since, not least because it happens to be particularly amenable to theoretical analysis. However, it has been recognised that there are some problems with this selection mechanism:

- Outstanding individuals take over the entire population very quickly. This tends to focus the search process, and makes it less likely that the algorithm will thoroughly search the space of possible solutions, where better solutions may exist. This phenomenon is often observed in early generations, when many of the randomly created individuals will have low fitness, and is known as **premature convergence**.

- When fitness values are all very close together, there is almost no **selection pressure**, so selection is almost uniformly random, and having a slightly better fitness is not very 'useful' to an individual. Therefore, later in a run, when some convergence has taken place and the worst individuals are gone, it is typically observed that the mean population fitness only increases very slowly.
- The mechanism behaves differently if the fitness function is transposed.

This last point is illustrated in Table 5.1, which shows three individuals and a fitness function with $f(A) = 1, f(B) = 4$, and $f(C) = 5$. Transposing this fitness function changes the selection probabilities, while the shape of the fitness landscape, and hence the location of the optimum, remains the same.

Individual	Fitness for f	Sel. prob. for f	Fitness for $f + 10$	Sel. prob. for $f + 10$	Fitness for $f + 100$	Sel. prob. for $f + 100$
A	1	0.1	11	0.275	101	0.326
B	4	0.4	14	0.35	104	0.335
C	5	0.5	15	0.375	105	0.339
Sum	10	1.0	40	1.0	310	1.0

Table 5.1. Transposing the fitness function changes selection probabilities for fitness-proportionate selection

To avoid the second two problems with FPS, a procedure known as **windowing** is often used. Under this scheme, fitness differentials are maintained by subtracted from the raw fitness $f(x)$ a value β^t, which depends in some way on the recent search history, and so can change over time (hence the superscript t). The simplest approach is just to subtract the value of the least-fit member of the current population P^t by setting $\beta^t = min_{y \in P^t} f(y)$ This value may fluctuate quite rapidly, so one alternative is to use a running average over the last few generations.

Another well-known approach is **sigma scaling** [189], which incorporates information about the mean $\bar{f}$ and standard deviation σ_f of fitnesses in the population:

$$f'(x) = max(f(x) - (\bar{f} - c \cdot \sigma_f), 0),$$

where c is a constant value, usually set to 2.

5.2.2 Ranking Selection

Rank-based selection is another method that was inspired by the observed drawbacks of fitness proportionate selection [32]. It preserves a constant selection pressure by sorting the population on the basis of fitness, and then allocating selection probabilities to individuals according to their rank, rather than according to their actual fitness values. Let us assume that the ranks are numbered so that an individual's rank notes how many worse solutions are in the population, so the best has rank μ-1 and the worst has rank 0. The mapping from rank number to selection probability can be done in many ways, for example, linearly or exponentially decreasing. As with FPS above, and any selection scheme, we insist that the sum over the population of the selection probabilities must be unity – that we must select *one* of the parents.

The usual formula for calculating the selection probability for linear ranking schemes is parameterised by a value s ($1 < s \leq 2$). In the case of a generational EA, where $\mu = \lambda$, this can be interpreted as the *expected* number of offspring allotted to the fittest individual. Since this individual has rank $\mu - 1$, and the worst has rank 0, then the selection probability for an individual of rank i is:

$$P_{lin-rank}(i) = \frac{(2-s)}{\mu} + \frac{2i(s-1)}{\mu(\mu-1)}.$$

Note that the first term will be constant for all individuals (it is there to ensure the probabilities add to one). Since the second term will be zero for the worst individual (with rank $i = 0$), it can be thought of as the 'baseline' probability of selecting that individual.

In Table 5.2 we show an example of how the selection probabilities differ for a population of $\mu = 3$ different individuals with fitness proportionate and rank-based selection with different values of s.

Individual	Fitness	Rank	P_{selFP}	P_{selLR} $(s = 2)$	P_{selLR} $(s = 1.5)$
A	1	0	0.1	0	0.167
B	4	1	0.4	0.33	0.33
C	5	2	0.5	0.67	0.5
Sum	10		1.0	1.0	1.0

Table 5.2. Fitness proportionate (FP) versus linear ranking (LR) selection

When the mapping from rank to selection probabilities is linear, only limited selection pressure can be applied. This arises from the assumption that, on average, an individual of median fitness should have one chance to be reproduced, which in turn imposes a maximum value of $s = 2$. (Since the scaling is linear, letting $s > 2$ would require the worst to have a negative selection probability if the probabilities are to sum to unity.) If a higher selection pressure is required, i.e., more emphasis on selecting individuals of above-average fitness, an exponential ranking scheme is often used, of the form:

$$P_{exp-rank}(i) = \frac{1 - e^{-i}}{c}.$$

The normalisation factor c is chosen so that the sum of the probabilities is unity, i.e., it is a function of the population size.

5.2.3 Implementing Selection Probabilities

The description above provides two alternative schemes for deciding a probability distribution that defines the likelihood of each individual in the population being selected for reproduction. In an ideal world, the mating pool of parents taking part in recombination would have exactly the same proportions as this selection probability distribution. This would mean that the number of any given individual would be given by its selection probability, multiplied by the size of the mating pool. However, in practice this is not possible because of the finite size of the population, i.e., when we do this multiplication, we find typically that some individuals have an *expected* number of copies which is noninteger – whereas of course in practice we need to select complete individuals. In other words, the mating pool of parents is *sampled* from the selection probability distribution, but will not in general accurately reflect it, as was seen in the example in Sect. 3.3.

The simplest way of achieving this sampling is known as the **roulette wheel** algorithm. Conceptually this is the same as repeatedly spinning a one-armed roulette wheel, where the sizes of the holes reflect the selection probabilities. In general, the algorithm can be applied to select λ members from the set of μ parents into a mating pool. To illustrate the workings of this algorithm, we will assume some order over the population (ranking or random) from 1 to μ, so that we can calculate the *cumulative probability distribution*, which is a list of values $[a_1, a_2, \ldots, a_\mu]$ such that $a_i = \sum_1^i P_{sel}(i)$, where $P_{sel}(i)$ is defined by the selection distribution — fitness proportionate or ranking. Note that this implies $a_\mu = 1$. The outlines of the algorithm are given in Fig. 5.1.

```
BEGIN
  /* Given the cumulative probability distribution a */
  /* and assuming we wish to select λ members of the mating pool */
  set current_member = 1;
  WHILE ( current_member ≤ λ ) DO
    Pick a random value r uniformly from [0,1];
    set i = 1;
    WHILE ( a_i < r ) DO
      set i = i + 1;
    OD
    set mating_pool[current_member] = parents[i];
    set current_member = current_member + 1;
  OD
END
```

Fig. 5.1. Pseudocode for the roulette wheel algorithm

Despite its inherent simplicity, it has been recognised that the roulette wheel algorithm does not in fact give a particularly good sample of the required distribution. Whenever more than one sample is to be drawn from the distribution – for instance λ – the use of the **stochastic universal sampling** (SUS) algorithm [32] is preferred. Conceptually, this is equivalent to making one spin of a wheel with λ equally spaced arms, rather than λ spins of a one-armed wheel. Given the same list of cumulative selection probabilities $[a_1, a_2, \ldots, a_\mu]$, it selects the mating pool as described in Fig. 5.2.

```
BEGIN
  /*  Given the cumulative probability distribution a */
  /*  and assuming we wish to select λ members of the mating pool */
  set current_member = i = 1;
  Pick a random value r uniformly from [0, 1/λ];
  WHILE ( current_member ≤ λ ) DO
    WHILE (  r ≤ a[i] ) DO
      set mating_pool[current_member] = parents[i];
      set r = r + 1/λ;
      set current_member = current_member + 1;
    OD
    set i = i + 1;
  OD
END
```

Fig. 5.2. Pseudocode for the stochastic universal sampling algorithm making λ selections

Since the value of the variable r is initialised in the range $[0, 1/\lambda]$ and increases by an amount $1/\lambda$ every time a selection is made, it is guaranteed that the number of copies made of each parent i is at least the integer part of $\lambda \cdot P_{sel}(i)$ and is no more than one greater. Finally, we should note that with minor changes to the code, SUS can be used to make any number of selections from the parents, and in the case of making just one selection, it is the same as the roulette wheel.

5.2.4 Tournament Selection

The previous two selection methods and the algorithms used to sample from their probability distributions relied on a knowledge of the entire population. However, in certain situations, for example, if the population size is very large, or if the population is distributed in some way (perhaps on a parallel system), obtaining this knowledge is either highly time consuming or at worst impossible. Furthermore, both methods assume that fitness is a quantifiable measure (based on some explicit objective function to be optimised), which may not be valid. Think, for instance, of an application evolving game playing strategies. In this case we might not be able to quantify the strength of a given individual (strategy) in isolation, but we can compare any two of them by simulating a game played by these strategies as opponents. Similar situations occur also in evolutionary design and evolutionary art applications [48, 49]. In these the user typically makes a subjective selection by comparing individuals representing designs or pieces of art, rather than using a quantitative measure to assign fitness, cf. Sect. 14.1.

Tournament selection is an operator with the useful property that it does not require any global knowledge of the population, nor a quantifiable measure of quality. Instead it only relies on an ordering relation that can compare and rank any two individuals. It is therefore conceptually simple and fast to implement and apply. The application of tournament selection to select λ members of a pool of μ individuals works according to the procedure shown in Fig. 5.3.

```
BEGIN
  /* Assume we wish to select λ members of a pool of μ individuals */
  set current_member = 1;
  WHILE ( current_member ≤ λ ) DO
    Pick k individuals randomly, with or without replacement;
    Compare these k individuals and select the best of them;
    Denote this individual as i;
    set mating_pool[current_member] = i;
    set current_member = current_member + 1;
  OD
END
```

Fig. 5.3. Pseudocode for the tournament selection algorithm

Because tournament selection looks at relative rather than absolute fitness, it has the same properties as ranking schemes in terms of invariance to translation and transposition of the fitness function. The probability that an individual will be selected as the result of a tournament depends on four factors, namely:

- Its rank in the population. Effectively this is estimated without the need for sorting the whole population.
- The **tournament size** k. The larger the tournament, the greater the chance that it will contain members of above-average fitness, and the less that it will consist entirely of low-fitness members. Thus the probability of selecting a high-fitness member increases, and that of selecting a low-

fitness member decreases, as k is increased. Hence we say that increasing k increases the selection pressure.

- The probability p that the most fit member of the tournament is selected. Usually this is 1 (*deterministic tournaments*), but stochastic versions are also used with $p < 1$. Since this makes it more likely that a less-fit member will be selected, decreasing p will decrease the selection pressure.
- Whether individuals are chosen with or without replacement. In the second case, with deterministic tournaments, the k-1 least-fit members of the population can never be selected, since the other member of the tournament will be fitter. However, if the tournament candidates are picked with replacement, it is always possible for even the least-fit member of the population to be selected, since with probability $1/\mu^k > 0$ all tournament candidates will be copies of that member.

These properties of tournament selection were characterised in [20, 58], and it was shown [190] that for binary ($k = 2$) tournaments with parameter p the expected time for a single individual of high fitness to take over the population is the same as that for linear ranking with $s = 2p$. However, since λ tournaments are required to produce λ selections, it suffers from the same problems as the roulette wheel algorithm, in that the outcomes can show a high variance from the theoretical probability distribution. Despite this drawback, tournament selection is perhaps the most widely used selection operator in some EC dialects (in particular, Genetic Algorithms), due to its extreme simplicity and the fact that the selection pressure is easy to control by varying the tournament size k.

5.2.5 Uniform Parent Selection

In some dialects of EC it is common to use mechanisms such that each individual has the same chance to be selected. At first sight this might appear to suggest that there is no selection pressure in the algorithm, which would indeed be true if this was not coupled with a strong fitness-based survivor selection mechanism.

In Evolutionary Programming, usually there is no recombination, only mutation, and parent selection is deterministic. In particular, each parent produces exactly one child by mutation. Evolution Strategies are also usually implemented with uniform random selection of parents into the mating pool, i.e., for each $1 \leq i \leq \mu$ we have $P_{uniform}(i) = 1/\mu$.

5.2.6 Overselection for Large Populations

In some cases it may be desirable to work with extremely large populations. Sometimes this could be for technical reasons – for example, there has been a lot of interest in implementing EAs using graphics cards (GPUs), which offer similar speed-up to clusters or supercomputers, but at much lower cost.

However, achieving the maximum potential speed-up typically depends on having a large population on each processing node.

Regardless of the implementation details, if the potential search space is enormous it might be a good idea to use a large population to avoid 'missing' promising regions in the initial random generation, and thereafter to maintain the diversity needed to support exploration. For example, in Genetic Programming it is not unusual to use population sizes of several thousands: in 1994 [254] used 1000; in 1996 [7] used 128,000; and in 1999 [255] used 1,120,000 individuals. In the latter case, often a method called **over-selection** is used for population sizes of 1000 and above.

In this method, the population is first ranked by fitness and then divided into two groups, the top x% in one and the remaining $(100-x)$% in the other. When parents are selected, 80% of the selection operations choose from the first group, and the other 20% from the second. Koza [252] provides rule of thumb values for x depending on the population size as shown in Table 5.3. As can be seen, the number of individuals from which the majority of parents are chosen stays constant, i.e., the selection pressure increases dramatically for larger populations.

Population size	Proportion of population in fitter group (x)
1000	32%
2000	16%
4000	8%
8000	4%

Table 5.3. Rule of thumb values for overselection: Proportion of ranked population in fitter subpopulation from which majority of parents are selected

5.3 Survivor Selection

The survivor selection mechanism is responsible for managing the process of reducing the working memory of the EA from a set of μ parents and λ offspring to a set of μ individuals forming the next generation. In principle, any of the mechanisms introduced for parent selection could be also used for selecting survivors. However, over the history of EC a number of special survivor selection strategies have been suggested and are widely used.

As explained in Sect. 3.2.6, this step in the main evolutionary cycle is also called replacement. In the present section we often use this latter term to be consistent with the literature. Replacement strategies can be categorised according to whether they discriminate on the basis of the fitness or the age of individuals.

5.3.1 Age-Based Replacement

The basis of these schemes is that the fitness of individuals is not taken into account during the selection of which individuals to replace in the population. Instead, they are designed so that each individual exists in the population for the same number of EA iterations. This does not preclude the possibly that *copies* of highly-fit individuals might persist in the population, but for this to happen they must be chosen at least once in the selection phase and then survive the recombination and mutation stages without being modified. Note that since fitness is not taken into account, the mean, and even best fitness of any given generation, may be lower than that of its predecessor. While slightly counterintuitive, this is not a problem as long as it does not happen too often, and may even be beneficial if the population is concentrated around a local optimum. A net increase in the mean fitness over time therefore relies on (i) having sufficient selection pressure when selecting parents into the mating pool, and (ii) using variation operators that are not too disruptive.

Age-based replacement is the strategy used in the simple Genetic Algorithm. Since the number of offspring produced is the same as the number of parents ($\mu = \lambda$), each individual exists for just one cycle, and the parents are simply discarded, to be replaced by the entire set of offspring. This is the generational model, but in fact this replacement strategy can also be implemented in a steady-state with overlapping populations ($\lambda < \mu$), right to the other extreme where a single offspring is created and inserted in the population in each cycle. In this case the strategy takes the form of a first-in-first-out (FIFO) queue.

An alternative method of age-based replacement for steady-state GAs is to randomly select a parent for replacement. A straightforward mathematical argument based on the population size being fixed tells us that this probabilistic strategy has the same mean effect – that is, *on average* individuals live for μ iterations. De Jong and Sarma [105] investigated this strategy experimentally, and found that the algorithm showed higher variance in performance than a comparable generational GA. Smith and Vavak [400] showed that this was because the random strategy is far more likely to lose the best member of the population than a delete-oldest (FIFO) strategy. For these reasons the random replacement strategy is not recommended.

5.3.2 Fitness-Based Replacement

A wide number of strategies based on fitness have been proposed for choosing which μ of the μ parents + λ offspring should go forward to the next generation. Some also take age into account.

Replace worst (GENITOR) In this scheme the worst λ members of the population are selected for replacement. Although this can lead to very rapid improvements in the mean population fitness, it can also lead to premature convergence as the population tends to rapidly focus on the fittest member currently present. For this reason it is commonly used in conjunction with large populations and/or a "no duplicates" policy.

Elitism This scheme is commonly used in conjunction with age-based and stochastic fitness-based replacement schemes, to prevent the loss of the current fittest member of the population. In essence a trace is kept of the current fittest member, and it is always kept in the population. Thus if it is chosen in the group to be replaced, and none of the offspring being inserted into the population has equal or better fitness, then it is kept and one of the offspring is discarded.

Round-robin tournament This mechanism was introduced within Evolutionary Programming, where it is applied to choose μ survivors. However, in principle, it can also be used to select λ parents from a given population of μ. The method works by holding pairwise tournament competitions in round-robin format, where each individual is evaluated against q others randomly chosen from the merged parent and offspring populations. For each comparison, a "win" is assigned if the individual is better than its opponent. After finishing all tournaments, the μ individuals with the greatest number of wins are selected. Typically, $q = 10$ is recommended in Evolutionary Programming. It is worth noting that this stochastic variant of selection allows for less-fit solutions to be selected if they had a lucky draw of opponents. As the value of q increases this chance becomes more and unlikely, until in the limit it becomes deterministic $\mu + \mu$.

$(\mu + \lambda)$ Selection The name and the notation of the $(\mu + \lambda)$ selection comes from Evolution Strategies. In general, it refers to the case where the set of offspring and parents are merged and ranked according to (estimated) fitness, then the top μ are kept to form the next generation. This strategy can be seen as a generalisation of the GENITOR method $(\mu > \lambda)$ and the round-robin tournament in Evolutionary Programming $(\mu = \lambda)$. In Evolution Strategies $\lambda > \mu$ with a great offspring surplus (typically $\lambda/\mu \approx 5 - 7$) that induces a large selection pressure.

(μ, λ) Selection The (μ, λ) strategy used in Evolution Strategies where typically $\lambda > \mu$ children are created from a population of μ parents. This method works on a mixture of age and fitness. The age component means that all the parents are discarded, so no individual is kept for more than one generation (although of course *copies* of it might exist later). The fitness component comes from the fact that the λ offspring are ranked according to the fitness, and the best μ form the next generation.

In Evolution Strategies, (μ, λ) selection, is generally preferred over $(\mu + \lambda)$ selection for the following reasons:

- The (μ, λ) discards all parents and is therefore in principle able to leave (small) local optima. This may be advantageous in a multimodal search space with many local optima.
- If the fitness function is not fixed, but changes in time, the $(\mu+\lambda)$ selection preserves outdated solutions, so it is not able to follow the moving optimum well.
- $(\mu + \lambda)$ selection hinders the self-adaptation mechanism used to adapt strategy parameters, cf. Sect. 6.2.

5.4 Selection Pressure

Throughout this chapter we have referred rather informally to the notion of selection pressure, using an intuitive description that as selection pressure increases, so fitter solutions are more likely to survive, or be chosen as parents, and less-fit solutions are correspondingly less likely.

A number of measures have been proposed for quantifying this, and studied theoretically, of which the best known is the takeover time. The **takeover time** τ^* of a given selection mechanism is defined as the number of generations it takes until the application of selection completely fills the population with copies of the best individual, given one copy initially. Goldberg and Deb [190] showed that

$$\tau^* = \frac{\ln \lambda}{\ln(\lambda/\mu)}.$$

For a typical evolution strategy with $\mu = 15$ and $\lambda = 100$, this results in $\tau^* \approx 2$. For fitness proportional selection in a genetic algorithm it is

$$\tau^* = \lambda \ln \lambda,$$

resulting in $\tau^* = 460$ for population size $\lambda = 100$.

Other authors have extended this analysis to other strategies in generational and steady-state population models [79, 400]; Rudolph applied it to different population structures such as rings [360], and also to consider a range of other measures of selection operators' performance, such as the 'Diversity Indicator'. Other measures of selection pressure have been proposed, including the 'Expected Loss of Diversity' [310], which is the expected change in the number of diverse solutions after μ selection events; and from theoretical biology, the 'Selection Intensity', which is the expected relative increase in mean population fitness after applying a selection operator.

While these measures can help in understanding the effect of different strategies, they can also be rather misleading since they consider selection alone, rather than in the context of variation operators providing diversity. Smith [390] derived mathematical expressions for a number of these indicators considering a wide range of replacement strategies in steady-state EAs. Experiments bore out the analytic results, and a benchmark comparison using well-known test problems showed that both the mean and variance of the takeover time could correctly predict the *relative* ordering of the mean and variance of the time taken to first locate the global optimum. However, for many applications of EAs the most important measure is the quality of the best solution found and also possibly the diversity of good solutions discovered. Smith's results showed that in fact none of the theoretical measures were particularly indicative of the relative performance of different algorithms in these terms.

5.5 Multimodal Problems, Selection, and the Need for Diversity

5.5.1 Multimodal Problems

In Sects. 2.3.1 and 3.5 we introduced the concept of multimodal search landscapes and local optima. We discussed how effective search relies on the preservation of sufficient diversity to allow both exploitation of learned information (by investigating regions contained high fitness solutions discovered) and exploration in order to uncover new high-fitness regions.

Multimodality is a typical aspect of the type of problems for which EAs are often employed, either in attempt to locate the global optimum (particularly when a local optimum has the largest basin of attraction), or to identify a number of high–fitness solutions corresponding to various local optima. The latter situation can often arise, for example, when the fitness function used by the EA does not completely specify the underlying problem. An example of this might be in the design of a new widget, where the parameters of the fitness function may change during the design process, as progressively more refined and detailed models are used as decisions such as the choice of materials, etc., are made. In this situation it is valuable to be able to examine a number of possible options, first so as to permit room for human aesthetic judgements, and second because it is probably desirable to use solutions from niches with broader peaks rather than from a sharp peak. This is because the latter may be overfitted (that is, overly specialised) to the current fitness function and may not be as good once the fitness function is refined.

The population-based nature of EAs holds out much promise for identifying multiple optima, however, in practice the finite population size, when coupled with recombination between *any* parents (known as **panmictic** mixing) leads to the phenomenon known as **genetic drift** and eventual convergence around one optimum. The reasons for this can easily be seen: imagine that we have two equally fit niches, and a population of 100 individuals originally equally divided between them. Eventually, because of the random effects in selection, it is likely that we will obtain a parent population consisting of 49 of one sort and 51 of the other. Ignoring the effects of recombination and mutation, in the next generation the probabilities of selecting individuals from the two niches are now 0.49 and 0.51 respectively, i.e., we are increasingly likely to select individuals from the second niche. This effect increases as the two subpopulations become unbalanced, until eventually we end up with only one niche represented in the population.

5.5.2 Characterising Selection and Population Management Approaches for Preserving Diversity

A number of mechanisms have been proposed to aid the use of EAs on multimodal problems. These can be broadly separated into two camps: *explicit* approaches, in which specific changes are made to operators in order to preserve diversity, and *implicit* approaches, in which a framework is used that permits, *but does not guarantee*, the preservation of diverse solutions. Before describing these it is useful to clarify exactly what we mean by 'diversity' and 'space'. Just as biological evolution takes place on a geographic surface, but can also be considered to occur on an adaptive landscape, so we can define a number of spaces within which the evolutionary algorithms operate:

- **Genotype Space:** We may perceive the set of representable solutions as a genotype space and define some distance metrics. This can be a natural distance metrics in that space (e.g., the Manhattan distance) or based on some fundamental move operator. Typical move operators include a single bit-flip for binary spaces, a single inversion for adjacency-based permutation problems and a single swap for order-based permutations problems.
- **Phenotype Space:** This is the end result: a search space whose structure is based on distance metrics between solutions. The neighbourhood structure in this space may bear little relationship to that in the genotype space according to the complexity of the representation–solution mapping.
- **Algorithmic Space:** This is the equivalent of the geographical space on which life on Earth has evolved. Effectively we are considering that the working memory of the EA, that is, the population of candidate solutions, can be structured in some way. This spatial structure could be either a conceptual division, or real: for example, a population might be split over a number of processors or cores.

Explicit approaches to diversity maintenance based on measures of either genotype or phenotypic space include Fitness Sharing (Sect. 5.5.3), Crowding (Sect. 5.5.4), and Speciation (Sect. 5.5.5), all of which work by affecting the probability distributions used by selection. Implicit approaches to diversity maintenance based on the concept of algorithmic space include Island Model EAs (Sect. 5.5.6) and Cellular EAs (Sect. 5.5.7).

5.5.3 Fitness Sharing

This scheme is based upon the idea that the number of individuals within a given niche is controlled by sharing their fitness immediately prior to selection, in an attempt to allocate individuals to niches *in proportion to the niche fitness* [193]. In practice the scheme considers each possible pairing of individuals i and j within the population (including i with itself) and calculates a distance $d(i,j)$ between them according to some distance metric (phenotypic is preferred if possible, else genotypic, e.g., Hamming distance for binary representations). The fitness F of each individual i is then adjusted according to the number of individuals falling within some prespecified distance σ_{share} using a power-law distribution:

$$F'(i) = \frac{F(i)}{\sum_j sh(d(i,j))},$$

where the sharing function $sh(d)$ is a function of the distance d, given by

$$sh(d) = \begin{cases} 1 - (d/\sigma_{share})^{\alpha} & \text{if } d \leq \sigma_{share}, \\ 0 & \text{otherwise}. \end{cases}$$

The constant value α determines the shape of the sharing function: for α=1 the function is linear, but for values greater than this the effect of similar individuals in reducing a solution's fitness falls off more rapidly with distance.

The value of the share radius σ_{share} decides both how many niches can be maintained and the granularity with which different niches can be discriminated. Deb [114] gives some suggestions for how this might be set if the number of niches is known *in advance*, but clearly this is not always the case. In [110] he suggests that a default value in the range 5–10 should be used.

We should point out that the use of fitness proportionate selection is implicit within the fitness-sharing method. In this case there exists a stable distribution of solutions amongst the niches when solutions from each peak have the same effective fitness F'. Since the niche fitness $F'_k = F_k/n_k$, in this stable distribution each niche k contains a number of solutions n_k proportional to the niche fitness F_k[1]. This point is illustrated in Fig. 5.4. Studies have indicated that the use of alternative selection methods does not lead to the formation and preservation of stable subpopulations in niches [324].

5.5.4 Crowding

The crowding algorithm was first suggested in De Jong's thesis [102] as a way of preserving diversity by ensuring that new individuals replaced *similar* members of the population. The original scheme worked in a steady-state setting (the number of new individuals generated in each step was 20% of the population size). When a new offspring is inserted into the population, a number [2] of members of the parent population are chosen at random, and then the offspring replaces the most similar of those parents. A number of problems

[1] This assumes for the sake of ease that all solutions within a given niche lie at its optimal point, at zero distance from each other.

[2] called the Crowding Factor (CF) - De Jong used CF=2

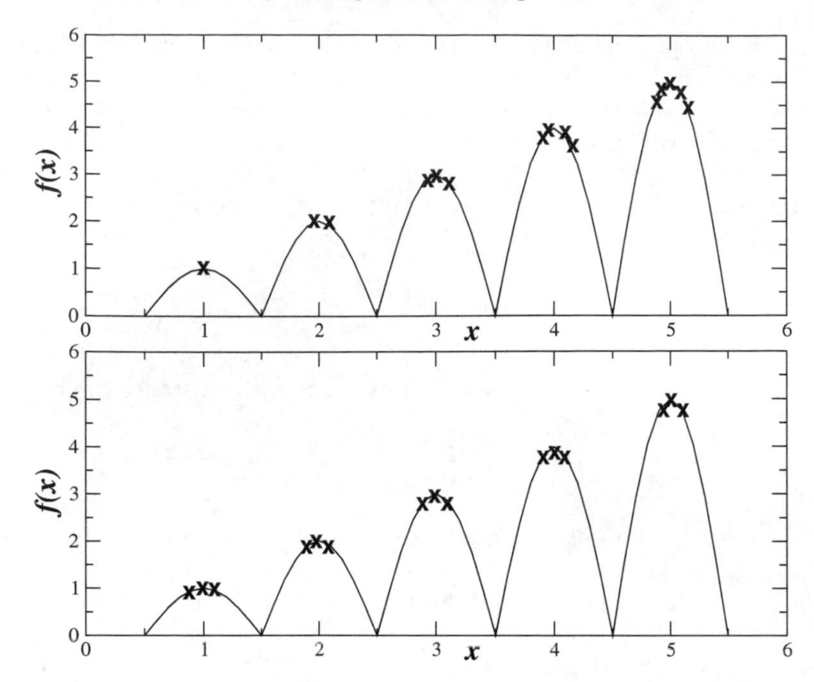

Fig. 5.4. Idealised population distributions under fitness sharing (top) and crowding (bottom). There are five peaks in the landscape with fitnesses (5,4,3,2,1) and the population size is 15. Fitness sharing allocates individuals to peaks in proportion to their fitness, whereas crowding distributes the population evenly amongst the peaks

were found with this approach, and Mahfoud has suggested an improvement called **deterministic crowding** [278]. This algorithm relies on the fact that offspring are likely to be similar to their parents as follows:

1. The parent population is randomly paired.
2. Each pair produces two offspring via recombination.
3. These offspring are mutated and then evaluated.
4. The four pairwise distances between offspring and parents are calculated.
5. Each offspring then competes for survival in a tournament with one parent, so that the intercompetition distances are minimised. In other words, denoting the parents as p, the offspring as o, and using the subscript to indicate tournament pairing, $d(p_1, o_1) + d(p_2, o_2) < d(p_1, o_2) + d(p_2, o_1)$.

The net result of all this is that offspring tend to compete for survival with the most similar parent, so subpopulations are preserved in niches but their size does not depend on fitness; rather it is equally distributed amongst the peaks available. Fig. 5.4 illustrates this point in comparison with the distribution achieved under crowding.

5.5.5 Automatic Speciation Using Mating Restrictions

The automatic speciation approach imposes mating restrictions based on some aspect of the candidate solutions (or their genotypes) defining them as belonging to different species. The population contains multiple species, and during parent selection for recombination individuals will only mate with others from the same (or similar) species. The biological analogy becomes particularly clear when we note that some authors refer to the aspect controlling reproductive opportunities as an individual's 'plumage' [401].

A number of schemes have been proposed to implement speciation, which can be divided into two main approaches. In the first speciation is based on the solution (or its representation), e.g., Deb's phenotype (genotype)-restricted mating [109, 114, 401]. The alternative approach is to add some elements such as tags to the genotype that code for the individual's species, rather than representing part of the solution. See [62, 109, 409] for implementations, noting that many of these ideas were previously suggested by other authors. These are usually randomly initialised and subject to recombination and mutation. Common to both approaches is the idea that once an individual has been selected to be a parent, then the choice of mate involves the use of a pairwise distance metric (in phenotype or genotype space as appropriate), with potential mates being rejected beyond a certain distance.

Note that in the tag scheme, there is initially no guarantee that individuals with similar tags will represent similar solutions, although after a few generations selection will usually take care of this problem. Neither is there any guarantee that different species will contain different solutions, although Spears goes some way towards rectifying this by also using the tags to perform fitness sharing [409], and even without this Deb reported improved performance compared to a standard GA [109]. Similarly, although the phenotype-based speciation scheme does not guarantee diversity maintenance, when used in conjunction with fitness sharing, it was reported to give better results than fitness sharing on its own [114].

5.5.6 Running Multiple Populations in Tandem: Island Model EAs

The idea of evolving multiple populations in tandem is also known as **island model EAs**, **parallel EA**, and, more precisely **coarse-grain parallel EAs**. These schemes attracted great interest in the 1980s when parallel computing became popular [87, 88, 274, 339, 356, 425] and are still applicable on MIMD systems such as computing clusters. Of course, they can equally well be implemented on a single-processor architecture, without the time speed-up.

The essential idea is to run multiple populations in parallel, in some kind of communication structure. The communication structure is usually a ring or a torus, but in principle any form is possible, and sometimes this is determined by the architecture of the parallel system, e.g., a hypercube [425]. After a (usually fixed) number of generations (known as an **epoch**), a number of individuals are selected from each population to be exchanged with others from neighbouring populations – this can be thought of as **migration**.

In [284] this approach is discussed in the context of Eldredge and Gould's theory of punctuated equilibria [154] and the exploration–exoloitation trade-off. They suggest that during the epochs between communication, when each subpopulation is evolving independently of the others, exploitation occurs, so that the subpopulations each explore the search space around the fitter solutions that they contain. When communication takes place, the injection of individuals of potentially high fitness, and with (possibly) radically different genotypes, facilitates exploration, particularly as recombination happens between the two different solutions.

Whilst extremely attractive in theory, it is obvious that there are no guarantees per se that the different subpopulations are actually exploring different regions of the search space. One possibility is clearly to achieve a start at this through a careful initialisation process, but even if this is used, there are a number of parameters that have been shown to affect the ability of this technique to explore different peaks and obtain good results even when only a single solution is desired as the end result.

A number of detailed studies have been made of the effects of different parameters and implementations of this basic scheme (see, e.g., earlier references in this section, and [276] for a more recent treatment), but of course we must bear in mind that the results obtained may be problem dependent, and so we will restrict ourselves to commenting on a few important facets:

- How often to exchange individuals? The essential problem here is that if the communication occurs too frequently, then all subpopulations will converge to the same solution. Equally if it is done too infrequently, and one or more subpopulations has converged quickly in the vicinity of a peak, then significant amounts of computational effort may be wasted. Most authors have used epoch lengths of the range 25–150 generations. An elegant alternative strategy proposed in [284] is to organise communication adaptively, that is to say, to stop the evolution in each subpopulation when no improvement has been observed for, say, 25 generations.
- How many, and which individuals to exchange? Many authors have found that in order to prevent too rapid convergence to the same solution, it is better to exchange a small number of solutions between subpopulations – usually 2–5. Once the amount of communication has been decided, it is necessary to specify *which* individuals are selected from each population to be exchanged. Clearly this can be done either by some fitness-based selection mechanism (e.g., "copy-best" [339], "pick-from-fittest-half" [425]) or at random [87]. It must also be decided whether the individuals being exchanged are effectively moved from one population to another, thus (assuming a symmetrical communication structure) maintaining subpopulation sizes, or whether they are merely copied, in which case each subpopulation must then undergo some kind of survivor selection mechanism.

The choices of how many and which individuals to exchange will evidently affect the tendency of the subpopulations to converge to the same solution. Random, rather than fitness-based, selection strategy is less likely to lead to takeover of one population by a new high-fitness migrant, and exchanging more solutions also leads to faster mixing and possible takeover. However, the extent to which these factors affect the behaviour is clearly tied to the epoch length, since if this is long enough to permit fitness convergence then all of the solutions contained within a given subpopulation are likely to be genotypically very similar, so the selection method used becomes less important.

- How to divide the population into subpopulations? The general rule here appears to be that provided a certain (problem-dependent) minimum subpopulation size is respected, then more subpopulations usually gives better results. This clearly fits in with our understanding, since if each subpopulation is exploring a different peak (the ideal scenario), the more peaks explored, the likely it is that one of them will contain the global optimum.

Finally, it is worth mentioning that it is perfectly possible to use different algorithmic parameters on different islands. Thus in the **injection island models** the subpopulations are arranged hierarchically with each level operating at a different granularity of representation. Equally, parameters such as the choice of recombination or mutation operator and associated parameters, or even subpopulation sizes, might be different between different subpopulations [148, 367].

5.5.7 Spatial Distribution Within One Population: Cellular EAs

In the previous section we described the implementation of a population structure in the form of a number of subpopulations with occasional communication. In this section we describe an alternative model whereby a single population is considered to be split into a larger number of smaller overlapping subpopulations (demes) by being distributed within algorithmic space. We can consider this to be equivalent to the situation whereby biological individuals are separated, only mating and competing for survival with those within a certain distance to them. To take a simple example from the days of less-rapid transport, a person might only have been able to marry and have children with someone from their own or surrounding villages. Thus should a new gene for say, telekinesis, evolve, even if it offers huge evolutionary advantage, at first it will only spread to surrounding villages. In the next generation it might spread to those surrounding them, and so on, only slowly diffusing or percolating throughout the society.

This effect is implemented by considering each member of the population to exist on a different point on a grid, and only permitting recombination and selection with neighbours, hence the common names of parallel EAs [195, 311], **fine-grain parallel EAs** [281], **diffusion model EA** [451], **distributed**

EAs [225] and, more commonly nowadays **cellular EAs** [456, 5]. There have been a great many differing implementations of this form of EA, but we can broadly outline the algorithm as follows:

1. The current population is conceptually distributed on a (usually toroidal) grid, with one individual per node.
2. For each node we have defined a deme (neighbourhood). This is usually the same for all nodes, e.g., for a neighbourhood size of nine on a square lattice, we take the node and all of its immediate neighbours.
3. In each generation we consider each deme in turn and perform the following operations within it:
   - Select two solutions from the nodes in the deme that will act as parents.
   - Generate an offspring via recombination.
   - Mutate, then evaluate the offspring.
   - Select one solution residing on a node in the deme and replace it with the new offspring.

Within this general structure there is scope for considerable differences in implementation. The ASPARAGOS algorithm [195, 311] uses a ladder topology rather than a lattice, and also performs a hill-climbing step after mutation. Several algorithms implemented on massively parallel SIMD or SPMD machines use asynchronous updates in step 3 rather than the sequential mode suggested in the third step above (a good discussion of this issue can be found in [338]). The selection of parents might be fitness-based [95] or random (or one of each [281]), and often one parent is taken to be that residing on the central node of the deme. When fitness-based selection is used it is usually a local implementation of a well-known global scheme such as fitness proportionate or tournament. De Jong and Sarma [106] analysed a number of such schemes and found that local selection techniques generally exhibited less selection pressure than their global versions. While it is common to replace the central node of the deme, again fitness-based or random selection have been used to select the individual to be replaced, or a combination such as "replace current solution if better" [195]. White and Pettey reported results suggesting that the use of fitness in the survivor selection is preferred [451]. A good recent treatment and discussion can be found in the book [5].

For exercises and recommended reading for this chapter, please visit `www.evolutionarycomputation.org`.

6

Popular Evolutionary Algorithm Variants

In this chapter we describe the most widely known evolutionary algorithm variants. This overview serves a twofold purpose: On the one hand, it introduces those historical EA variants without which no EC textbook would be complete together with some more recent versions that deserve their own place in the family tableau. On the other hand, it demonstrates the diversity of realisations of the same basic evolutionary algorithm concept.

6.1 Genetic Algorithms

The genetic algorithm (GA) is the most widely known type of evolutionary algorithm. It was initially conceived by Holland as a means of studying adaptive behaviour, as suggested by the title of the book describing his early research: *Adaptation in Natural and Artificial Systems* [220]. However, GAs have largely (if perhaps mistakenly – see [103]) been considered as function optimisation methods. This is perhaps partly due to the title of Goldberg's seminal book: *Genetic Algorithms in Search, Optimization and Machine Learning* [189] and some very high-profile early successes in solving optimisation problems. Together with De Jong's thesis [102] this work helped to define what has come to be considered as the classical genetic algorithm — commonly referred to as the 'canonical' or 'simple GA' (SGA). This has a binary representation, fitness proportionate selection, a low probability of mutation, and an emphasis on genetically inspired recombination as a means of generating new candidate solutions. It is summarised in Table 6.1. Perhaps because it is so widely used for teaching EAs, and is the first EA that many people encounter, it is worth re-iterating that many features that have been developed over the years are missing from the SGA — most obviously that of elitism.

While, the table does not indicate this, GAs traditionally have a fixed workflow: given a population of μ individuals, parent selection fills an intermediary population of μ, allowing duplicates. Then the intermediary population is shuffled to create random pairs and crossover is applied to each consecutive pair with probability p_c and the children replace the parents immediately. The new intermediary population undergoes mutation individual by individual, where each of the l bits in an individual is modified by mutation with independent probability p_m. The resulting intermediary population forms the next generation replacing the previous one entirely. Note that in this new generation there might be pieces, perhaps complete individuals, from the previous one that survived crossover and mutation without being modified, but the likelihood of this is rather low (depending on the parameters μ, p_c, p_m).

Representation	Bit-strings
Recombination	1-Point crossover
Mutation	Bit flip
Parent selection	Fitness proportional - implemented by Roulette Wheel
Survival selection	Generational

Table 6.1. Sketch of the simple GA

In the early years of the field there was significant attention paid to trying to establish suitable values for GA parameters such as the population size, crossover and mutation probabilities. Recommendations were for mutation rates between $1/l$ and $1/\mu$, crossover probabilities around 0.6-0.8, and population sizes in the fifties or low hundreds, although to some extent these values reflect the computing power available in the 1980s and 1990s.

More recently it has been recognised that there are some flaws in the SGA. Factors such as elitism, and non-generational models were added to offer faster convergence if needed. As discussed in Chap. 5, SUS is preferred to roulette wheel implementation, and most commonly rank-based selection is used, implemented via tournament selection for simplicity and speed. Studying the biases in the interplay between representation and one-point crossover (e.g. [411]) led to the development of alternatives such as uniform crossover, and a stream of work through 'messy-GAs' [191] and 'Linkage Learning' [209, 395, 385, 83] to Estimation of Distribution Algorithms (see Sect. 6.8). Analysis and experience has recognised the need to use non-binary representations where more appropriate (as discussed in Chap. 4). Finally the problem of how to choose a suitable fixed mutation rate has largely been solved by adopting the idea of self-adaptation, where the rates are encoded as extra genes in an individuals representation and allowed to evolve [18, 17, 396, 383, 375].

Nevertheless, despite its simplicity, the SGA is still widely used, not just for teaching purposes, and for benchmarking new algorithms, but also for relatively straightforward problems in which binary representation is suitable. It has also been extensively modelled by theorists (see Chap. 16). Since it has provided so much inspiration and insight into the behaviour of evolutionary processes in combinatorial search spaces, it is fair to consider that if OneMax is the *Drosophilia* of combinatorial problems for researchers, then the SGA is the *Drosophilia* of evolutionary algorithms.

6.2 Evolution Strategies

Evolution strategies (ES) were invented in the early 1960s by Rechenberg and Schwefel, who were working at the Technical University of Berlin on an application concerning shape optimisation (see [54] for a brief history). The earliest ES's were simple two-membered algorithms denoted (1+1) ES's (pronounce: one plus one ES), working in a vector space. An offspring is generated by the addition of a random number independently to each to the elements of the parent vector and accepted if fitter. An alternative scheme, denoted as (1,1) ES (pronounce: one comma one ES) always replaces the parent by the offspring, thus forgetting the previous solutions by definition. The random numbers are drawn from a Gaussian distribution with mean zero and a standard deviation σ, where σ is called the mutation step size. One of the key early breakthroughs of ES research was to propose a simple mechanism for online adjustment of step sizes by the famous **1/5 success rule** of Rechenberg [352] as described in Sect. 8.2.1. In the 1970s the concept of multi-membered evolution strategies was introduced, with the naming convention based on μ individuals in the population and λ offspring generated in one cycle. The resulting $(\mu+\lambda)$ and (μ,λ) ES's gave rise to the possibility of more sophisticated forms of step-size control, and led to the development of a very useful feature in evolutionary computing: **self-adaptation** of strategy parameters, see Sect. 4.4.2. In general, self-adaptivity means that some parameters of the EA are varied during a run in a specific manner: the parameters are included in the chromosomes and coevolve with the solutions. Technically this means that an ES works with extended chromosomes $\langle \overline{x}, \overline{p} \rangle$, where $\overline{x} \in \mathbb{R}^n$ is a vector from the domain of the given objective function to be optimised, while $\overline{p}$ carries the algorithm parameters. Modern evolution strategies always self-adapt the mutation step sizes and sometimes their rotation angles. That is, since the procedure was detailed in 1977 [372] most ESs have been self-adaptive, and other EAs have increasingly adopted self-adaptivity. Recent forms of ES such as the CMA [207] are among the leading algorithms for optimisation of complex real-valued functions. A summary of ES is given in Table 6.2.

Representation	Real-valued vectors
Recombination	Discrete or intermediary
Mutation	Gaussian perturbation
Parent selection	Uniform random
Survivor selection	Deterministic elitist replacement by (μ,λ) or $(\mu+\lambda)$
Speciality	Self-adaptation of mutation step sizes

Table 6.2. Sketch of ES

The basic recombination scheme in evolution strategies involves two parents that create one child. To obtain λ offspring recombination is performed λ

times. There are two recombination variants distinguished by the manner of recombining parent alleles. Using **discrete recombination** one of the parent alleles is randomly chosen with equal chance for either parents. In **intermediate recombination** the values of the parent alleles are averaged. An extension of this scheme allows the use of more than two recombinants, because the two parents are drawn randomly for each position $i \in \{1, \ldots, n\}$ in the offspring anew. These drawings take the whole population of μ individuals into consideration, and the result is a recombination operator with possibly more than two individuals contributing to the offspring. The exact number of parents, however, cannot be defined in advance. This multiparent variant is called **global recombination**. To make terminology unambiguous, the original variant is called **local recombination**. Evolution strategies typically use global recombination. Interestingly, different recombination is used for the object variable part (discrete is recommended) and the strategy parameters part (intermediary is recommended). This scheme preserves diversity within the phenotype (solution) space, allowing the trial of very different combinations of values, whilst the averaging effect of intermediate recombination assures a more cautious adaptation of strategy parameters.

The selection scheme that is generally used in evolution strategies is (μ, λ) selection, which is preferred over $(\mu + \lambda)$ selection for the following reasons:

- The (μ, λ) discards all parents and so can in principle leave (small) local optima, which is advantageous for multimodal problems.
- If the fitness function changes over time, the $(\mu + \lambda)$ selection preserves outdated solutions, so is less able to follow the moving optimum.
- $(\mu + \lambda)$ selection hinders the self-adaptation, because misadapted strategy parameters may survive for a relatively large number of generations. For example, if an individual has relatively good object variables but poor strategy parameters, often all of its children will be bad. Thus they will be removed by an elitist policy, while the misadapted strategy parameters in the parent may survive for longer than desirable.

The selective pressure in evolution strategies is very high because λ is typically much higher than μ (traditionally a 1/7 ratio is recommended, although recently values around 1/4 seem to gain popularity). The **takeover time** τ^* of a given selection mechanism is defined as the number of generations it takes until the application of selection completely fills the population with copies of the best individual, given one copy initially. Goldberg and Deb [190] showed that

$$\tau^* = \frac{\ln \lambda}{\ln(\lambda/\mu)}.$$

For a typical evolution strategy with $\mu = 15$ and $\lambda = 100$, this results in $\tau^* \approx 2$. By way of contrast, for fitness proportional selection in a genetic algorithm with $\mu = \lambda = 100$ it is $\tau^* = \lambda \ln \lambda = 460$. This indicates that an ES is a more aggressive optimizer than a (simple) GA.

6.3 Evolutionary Programming

Evolutionary programming (EP) was originally developed by Fogel et al. in the 1960s to simulate evolution as a learning process with the aim of generating artificial intelligence [166, 174]. Intelligence, in turn, was viewed as the capability of a system to adapt its behaviour in order to meet some specified goals in a range of environments. Adaptive behaviour is the key term in this definition, and the capability to predict the environment was considered to be a prerequisite. The classic EP systems used finite state machines as individuals.

Nowadays EP frequently uses real-valued representations, and so has almost merged with ES. The principal differences lie perhaps in the biological inspiration: in EP each individual is seen as corresponding to a distinct *species*, and so there is no recombination. Furthermore, the selection mechanisms are different. In ES parents are selected stochastically, then the selection of the μ best from the union of $\mu + \lambda$ offspring is deterministic. By contrast, in EP each parent generates exactly one offspring (i.e., $\lambda = \mu$), but these parents and offspring populations are then merged and compete in stochastic round-robin tournaments for survival, as described in Sect. 5.3.2. The field now adopts a very open, pragmatic approach that the choice of representation, and hence mutation, should be driven by the problem; Table 6.3 is therefore a representative rather than a standard algorithm variant.

Representation	Real-valued vectors
Recombination	None
Mutation	Gaussian perturbation
Parent selection	Deterministic (each parent creates one offspring via mutation)
Survivor selection	Probabilistic ($\mu + \mu$)
Speciality	Self-adaptation of mutation step sizes (in meta-EP)

Table 6.3. Sketch of EP

The issue of the advantage of using a mutation-only algorithm versus a recombination and mutation variant has been intensively discussed since the 1990s. Fogel and Atmar [170] compared the results of EP algorithms with and without recombination on a series of linear functions with parameterisable interactions between genes. They concluded that improved performance was obtained from the version without recombination. This led to intensive periods of research in both the EP and the GA communities to try and establish the circumstances under which the availability of a recombination operator yielded improved performance [159, 171, 222, 408]. The current state of thinking has moved on to a stable middle ground. The latest results [232] confirm that the ability of both crossover or Gaussian mutation to produce new offspring of superior fitness to their parents depends greatly on the state of the search process, with mutation better initially but crossover gaining in ability as evolution progresses. These conclusions agree with theoretical results developed elsewhere and discussed in more depth in Chap. 7. In particular it is stated that: "the traditional practice of setting operator probabilities at constant values, ... is quite limiting and may even prevent the successful discovery of suitable solutions." However, it is perhaps worth noting that even in these studies the authors did not detect a difference between the performance of different crossover operators, which they claim casts significant doubt on the building block hypothesis (Sect. 16.1), so we are not entirely without healthy scientific debate!

Since the 1990s EP variants for optimisation of real-valued parameter vectors have become more frequent and even positioned as 'standard' EP [22, 30]. During the history of EP a number of mutation schemes, such as one in which the step size is inversely related to the fitness of the solutions, have been proposed. Since the proposal of **meta-EP** [165, 166], self-adaptation of step sizes has become the norm, using the scheme in Eq. (4.4). A variety of schemes have been proposed, including mutation variables first then strategy parameters (which violates the rationale explained in Sect. 4.4.2). Tracing the literature on this issue, the paper by Gehlhaar and Fogel [182] seems to be a turning point. Here the authors explicitly compare the 'sigma first' and 'sigma last' strategies and conclude that the first one – the standard ES manner – offers a consistent general advantage over the second one. Notably in a paper and book [81, 168], Fogel uses the lognormal adaptation of n standard deviations σ_i, *followed* by the mutation of the object variables x_i themselves, suggesting that EP is practically merging with ES regarding this aspect. Other ideas from ES have also informed the development of EP algorithms, and a version with self-adaptation of covariance matrices, called **R-meta-EP** is also in use. Worthy of note is Yao's improved fast evolutionary programming algorithm (IFEP) [470], whereby two offspring are created from each parent, one using a Gaussian distribution to generate the random mutations, and the other using the Cauchy distribution. The latter has a fatter tail (i.e., more chance of generating a large mutation), which the authors suggest gives the overall algorithm greater chance of escaping from local minima, whilst the Gaussian distribution (if small step sizes evolve) gives greater ability to fine-tune the current parents.

6.4 Genetic Programming

Genetic programming is a relatively young member of the evolutionary algorithm family. It differs from other EA strands in its application area as well as the particular representation (using trees as chromosomes). While the EAs discussed so far are typically applied to optimisation problems, GP could instead be positioned in machine learning. In terms of the different problem types as discussed in Chapter 2, most other EAs are for finding some input realising maximum payoff (Fig. 1.1), whereas GP is used to seek models with maximum fit (Fig. 1.2). Clearly, once maximisation is introduced, modelling problems can be seen as special cases of optimisation. This, in fact, is the basis of using evolution for such tasks: models — represented as parse trees — are treated as individuals, and their fitness is the model quality to be maximised. The summary of GP is given in Table 6.4.

Representation	Tree structures
Recombination	Exchange of subtrees
Mutation	Random change in trees
Parent selection	Fitness proportional
Survivor selection	Generational replacement

Table 6.4. Sketch of GP

The parse trees used by GP as chromosomes capture expressions in a given formal syntax. Depending on the problem at hand, and the users' perceptions on what the solutions must look like, this can be the syntax of arithmetic expressions, formulas in first-order predicate logic, or code written in a programming language, cf. Sect. 4.6. In particular, they can be envisioned as executable codes, that is, programs. The syntax of functional programming, e.g., the language LISP, very closely matches the so-called Polish notation of expressions. For instance, the formula in Eq. (4.8) can be rewritten in this Polish notation as

$$+(\cdot(2,\pi),-(+(x,3),/(y,+(5,1)))),$$

while the executable LISP code[1] looks like:

$$(+\ (\cdot\ 2\ \pi)\ (-\ (+\ x\ 3)\ (/\ y\ (+\ 5\ 1)))).$$

Based on this perception, GP can be positioned as the "programming of computers by means of natural selection" [252], or the "automatic evolution of computer programs" [37].

There are a few other issues that are specific to tree-based representations, and hence (but not exclusively) to genetic programming.

Initialisation can be carried out in different ways for trees. The most common method used in GP is the so-called **ramped half-and-half** method. In this method a maximum initial depth D_{max} of trees is chosen, and then each member of the initial population is created from the sets of functions F and terminals T using one of the two methods below with equal probability:

- *Full method*: here each branch of the tree has depth D_{max}. The contents of nodes at depth d are chosen from F if $d < D_{max}$ or from T if $d = D_{max}$.

[1] To be precise we should use PLUS, etc., for the operators.

- *Grow method*: here the branches of the tree may have different depths, up to the limit D_{max}. The tree is constructed beginning from the root, with the contents of a node being chosen stochastically from $F \cup T$ if $d < D_{max}$.

Stochastic Choice of a Single Variation Operator. In GP offspring are typically created by either recombination *or* mutation, rather than recombination *followed by* mutation, as is more common in other variants. This difference is illustrated in Fig. 6.1 (inspired by Koza [252]), which compares the loop for filling the next generation in a generational GA with that of GP.

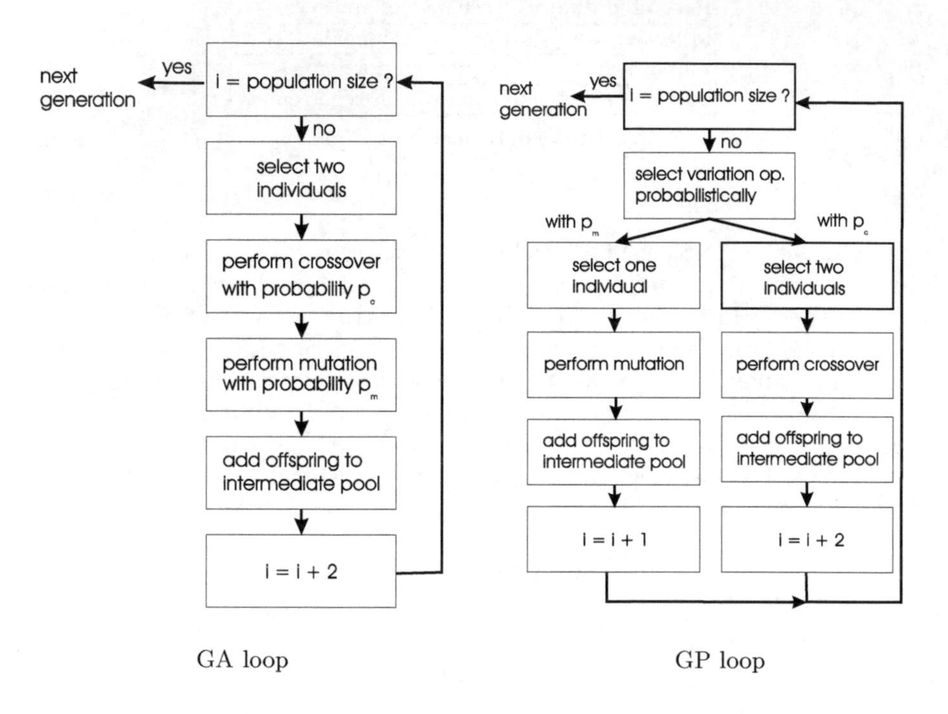

Fig. 6.1. GP flowchart versus GA flowchart. The two diagrams show two options for filling the intermediary population in a generational scheme. In a conventional GA mutation and crossover are used to produce the next offspring (left). Within GP, a new individual is created by either mutation or crossover (right).

Low or Zero Mutation Probabilities. Koza's classic book on GP from 1992 [252] advises users to set the mutation rate at 0, i.e., it suggests that GP works *without* mutation. More recently Banzhaf et al. recommended 5% [37]. In giving mutation such a limited role, GP differs from other EA streams. The reason for this is the generally shared view that crossover has a large shuffling effect, acting in some sense as a macromutation operator [9]. The current GP practice uses low mutation frequencies, even though some studies indicate that an (almost) pure crossover approach might be inferior [275].

Over-selection is often used to deal with the typically large population sizes (population sizes of several thousands are not unusual in GP). The method first ranks the population, then divides it into two groups, one containing the top $x\%$ and the other containing the other $(100-x)\%$. When parents are selected, 80% of the selection operations come from the first group, and the other 20% from the second group. The values of x used are found empirically by rule of thumb and depend on the population size with the aim that the number of individuals from which the majority of parents are chosen stays constant in the low hundreds, i.e., the selection pressure increases dramatically for larger populations.

Bloat (sometimes called the 'survival of the fattest') is a phenomenon observed in GP whereby average tree sizes tend to grow during the course of a run. There are many studies devoted to understanding why bloat occurs and to proposing countermeasures, see for instance [268, 407]. Although the results and discussions are not conclusive, one primary suspect is the sheer fact that we have chromosomes with variable length, meaning that the *possibility* for chromosome sizes to grow along the evolution already implies that they will *actually* do so. Probably the simplest way to prevent bloat is to introduce a maximum tree size and forbid a variation operator if the child(ren) resulting from its application would exceed this maximum size. In this case, this threshold can be seen as an additional parameter of mutation and recombination in GP. Several advanced techniques have also been proposed, but the only one that is widely acknowledged is that of **parsimony pressure**. Such a pressure towards parsimony (i.e., being 'stingy' or ungenerous) is achieved through introducing a penalty term in the fitness formula that reduces the fitness of large chromosomes [228, 406] or using multiobjective techniques [115].

6.5 Learning Classifier Systems

Learning Classifier Systems (LCS) represent an alternative evolutionary approach to model building based on the use of rule sets, rather than parse trees, to represent knowledge [270, 269]. LCS are used primarily in applications where the objective is to evolve a system that will respond to the current state of its environment (i.e., the inputs to the system) by suggesting a response that in some way maximises future reward from the environment.

An LCS is therefore a combination of a classifier system and a learning algorithm. The classifier system component is typically a set of rules, each mapping certain inputs to actions. The whole rule set therefore constitutes a model that covers the space of possible inputs and suggests the most appropriate actions for each. The learning algorithm component of an LCS is implemented by an evolutionary algorithm, whose population members either represent individual rules, or complete rule sets, known respectively as the Michigan and Pittsburgh approaches. The fitness driving the evolutionary process may be driven by many different forms of learning, here we restrict ourselves to 'supervised' learning, where at each stage the system receives a training signal (reward) from the environment in response to the output it proposes. This helps emphasise the difference between the Michigan and Pittsburgh approaches. In the former, data items are presented to the system one-by-one and individual rules are rewarded according to their predictions. By contrast, in a Pittsburgh approach each individual represents a complete model, so the fitness would normally be calculated by presenting the entire data set and calculating the mean accuracy of the predictions.

The **Michigan-style LCS** was first described by Holland in 1976 as a framework for studying learning in condition/action rule-based systems, using genetic algorithms as the principal method for the discovery of new rules and the reinforcement of successful ones [219]. Typically each member of the population was a single rule representing a partial model – that is to say it might only cover a region of the decision space. Thus it is the entire population that together represents the learned model. Each rule is a tuple {`condition:action:payoff`}. The condition specifies a region of the space of possible inputs in which the rule applies. The condition parts of rules may contain wildcard, or 'don't-care' characters for certain variables, or may describe a set of values that a given variable may take – for example, a range of values for a continuous variable. Rules may be distinguished by the number of wildcards they contain, and one rule is said to be more specific than another if it contains fewer wildcards, or if the ranges for certain variables are smaller — in other words if it covers a smaller region of the input space. Given this flexibility, it is common for the condition parts of rules to overlap, so a given input may match a number of rules. In the terminology of LCS, the subset of rules whose condition matches the current inputs from the environment is known as the **match set**. These rules may prescribe different actions, of which one is chosen. The action specifies either the action to be taken (for example, if controlling robots or on-line trading agents) or the system's prediction (such as a class label or a numerical value). The subset of the match set advocating the chosen action is known as the **action set**. Holland's original framework maintained lists of which rules have been used, and when a reward was received from the environment a portion was passed back to recently used rules to provide information for the selection mechanism. The intended effect is that the strength of a rule predicts the value of the reward that the system will gain for undertaking the action. However the framework proved unwieldy and difficult to make work well in practice.

LCS research was reinvigorated in the mid-1990s by Wilson who removed the concept of memory and stripped out all but the essential components in his minimalist ZCS algorithm [464]. At the same time several authors were noting the conceptual similarity between LCS and reinforcement learning algorithms which attempt to learn, for each input state, an accurate mapping from possible actions to expected rewards. The XCS algorithm [465] firmly established this link by extending rule-tuples to {`condition:action:payoff,accuracy`}, where the accuracy value reflects the system's experience of how well the predicted payoff matches the reward received. Unlike ZCS, the EA is restricted at each cycle — originally to the match set, latterly to the action set, which increases the pressure to discover generalised conditions for each action. As per ZCS, a credit assignment mechanism is triggered by the receipt of rewards from the environment to update the predicted pay-offs for rules in the previous action set. However, the major difference is that these are not used directly to drive selection in the evolution process. Instead selection operates on the basis of accuracy, so the algorithm can in principle evolve a *complete* mapping from input space to actions.

Table 6.5 gives a simple overview of the major features of Michigan-style classifiers for a problem with a binary input and output space. The list below summarizes the main workflow of the algorithm.

1. A new set of inputs are received from the environment.
2. The rule base is examined to find the match-set of rules.
   - If the match set is empty, a 'cover operator' is invoked to generate one or more new matching rules with a random action.
3. The rules in the match-set are grouped according to their actions.
4. For each of these groups the mean accuracy of the rules is calculated.
5. An action is chosen, and its corresponding group noted as the action set.
   - If the system is an 'exploit' cycle, the action with the highest mean accuracy is chosen.
   - If the system is in an 'explore' cycle, an action is chosen randomly or via fitness-proportionate selection, acting on the mean accuracies.
6. The action is carried out and a reward is received from the environment.
7. The estimated accuracy and predicted payoffs are then updated for the rule in the current and previous action sets, based on the rewards received and the predicted pay-offs, using a Widrow–Hoff style update mechanism.
8. If the system is in an 'explore' cycle, an EA is run within the action-set, creating new rules (with pay-off and accuracies set to the mean of their parents), and deleting others.

Representation	tuple of {`condition:action:payoff,accuracy`} conditions use {0,1,#} alphabet
Recombination	One-point crossover on conditions/actions
Mutation	Binary/ternary resetting as appropriate on action/conditions
Parent selection	Fitness proportional with sharing within environmental niches
Survivor selection	Stochastic, inversely related to number of rules covering same environmental niche
Fitness	Each reward received updates predicted payoff and accuracy of rules in relevant action sets by reinforcement learning.

Table 6.5. Sketch of a Michigan-style LCS for a binary input and action space

The **Pittsburgh-style LCS** predates, but is similar to the better-known GP: each member of the evolutionary algorithm's population represents a complete model of the mapping from input to output spaces. Each gene in an individual typically represents a rule, and again a new input item may match more than one rule, in which case typically the first match is taken. This means that the representation should be viewed as an ordered list, and two individuals which contain the same rules, but in a different order on the genome, are effectively different models. Learning of appropriately complex models is typically facilitated by using a variable-length representation so that new rules can be added at any stage. This approach has several conceptual advantages — in particular, since fitness is awarded to complete rule sets, models can be learned for complex multi-step problems. The downside of this flexibility is that, like GP, Pittsburgh-style LCS suffers from bloat and the search space becomes potentially infinite. Nevertheless, given sufficient computational resources, and effective methods of parsimony to counteract bloat, Pittsburgh-style LCS has demonstrated state-of-the-art performance in several machine learning domains, especially for applications such as bio-informatics and medicine, where human-interpretability of the evolved models is vital and large data-sets are available so that the system can evolve off-line to minimise prediction error. Two recent examples winning Humies Awards for better than human performance are in the realms of prostate cancer detection [272] and protein structure prediction [16].

6.6 Differential Evolution

In this section we describe a young, but powerful member of the evolutionary algorithm family: differential evolution (DE). Its birth can be dated to 1995, when Storn and Price published a technical report describing the main concepts behind a "new heuristic approach for minimizing possibly nonlinear and nondifferentiable continuous space functions" [419]. The distinguishing feature that delivered the name of this approach is a twist to the usual reproduction operators in EC: the so-called **differential mutation**. Given a population of candidate solution vectors in $\mathbb{R}^n$ a new mutant vector $\overline{x}'$ is produced by adding a **perturbation vector** to an existing one,

$$\overline{x}' = \overline{x} + \overline{p},$$

where the perturbation vector $\overline{p}$ is the scaled vector difference of two other, randomly chosen population members

$$\overline{p} = F \cdot (\overline{y} - \overline{z}), \tag{6.1}$$

and the scaling factor $F > 0$ is a real number that controls the rate at which the population evolves. The other reproduction operator is the usual uniform crossover, subject to one parameter, the crossover probability $Cr \in [0, 1]$ that defines the chance that for any position in the parents currently undergoing crossover, the allele of the first parent will be included in the child. (Remember that in GAs the crossover rate $p_c \in [0, 1]$ is defined for any given pair of individuals and it is the likelihood of actually executing crossover.) DE also has a slight twist to the crossover operator: at one randomly chosen position the child allele is taken from the first parent without making a random decision. This ensures that the child does not duplicate the second parent.

In the main DE workflow populations are lists, rather than (multi)sets, allowing references to the i-th individual by its position $i \in \{1, \ldots, \mu\}$ in this list. The order of individuals in such a population $P = \langle \overline{x}_1, \ldots, \overline{x}_i, \ldots, \overline{x}_\mu \rangle$ is not related to their fitness values. An evolutionary cycle starts with creating a mutant vector population $M = \langle \overline{v}_1, \ldots, \overline{v}_\mu \rangle$. For each new mutant $\overline{v}_i$ three vectors are chosen randomly from P, a base vector to be mutated and two others to define a perturbation vector. After making the mutant vector population, a so-called trial vector population $T = \langle \overline{u}_1, \ldots, \overline{u}_\mu \rangle$ is created, where $\overline{u}_i$ is the result of applying crossover to $\overline{v}_i$ and $\overline{x}_i$. (Note, that it is guaranteed that $\overline{u}_i$ does not duplicate $\overline{x}_i$.) In the last step deterministic selection is applied to each pair $\overline{x}_i$ and $\overline{u}_i$: the i-th individual in the next generation is $\overline{u}_i$ if $f(\overline{u}_i) \leq f(\overline{x}_i)$ and $\overline{x}_i$ otherwise.

Representation	Real-valued vectors
Recombination	Uniform crossover
Mutation	Differential mutation
Parent selection	Uniform random selection of the 3 necessary vectors
Survival selection	Deterministic elitist replacement (parent vs. child)

Table 6.6. Sketch of differential evolution

In general, a DE algorithm has three parameters, the scaling factor F, the population size μ (usually denoted by *NP* in the DE literature), and the crossover probability Cr. It is worth noting that despite mediating a crossover process, Cr can also be thought of as a mutation rate, i.e., the approximate probability that an allele will be inherited from a mutant [343]. The DE community also emphasises another aspect of uniform crossover: The number of inherited mutant alleles follows a binomial distribution, since allele origins are determined by a finite number of independent trials having two outcomes with constant probabilities.

Over the years, several DE variants have been invented and published. One of the modifications concerns the choice of the base vector when building the mutant population M. It can be randomly chosen for each v_i, as presented here, but it can be fixed, always using the best vector in the population and only varying the perturbation vectors. Another extension is obtained by allowing more than one difference vector to define the perturbation vector in the mutation operator. For example, using two difference vectors, Equation

6.1 becomes

$$\overline{p} = F \cdot (\overline{y} - \overline{z} + \overline{y}' - \overline{z}') \tag{6.2}$$

where $\overline{y}, \overline{z}, \overline{y}', \overline{z}'$ are four randomly chosen population members.

In order to classify the different variants, the notation DE/a/b/c has been introduced in the literature, where a specifies the base vector, e.g., "rand" or "best", b is the number of difference vectors used to define the perturbation vector, and c denotes the crossover scheme, e.g., "bin" stands for using uniform crossover (because of the binomial distribution of donor alleles it generates). Using this notation, the basic version described above is DE/rand/1/bin.

6.7 Particle Swarm Optimisation

The algorithm we describe here deviates somewhat from other evolutionary algorithms in that it is inspired by social behavior of bird flocking or fish schooling, while the name and the technical terminology are grounded in physical particles [340, 248]. Seemingly there is no evolution in a particle swarm optimizer, but algorithmically it does fit in the general EA framework. Particle swarm optimisation (PSO) was launched in 1995, when Kennedy and Eberhart published their seminal paper about a "concept for the optimization of nonlinear functions using particle swarm methodology" [247]. Similarly to DE, the distinguishing feature of PSO is a twist to the usual reproduction operators in EC: PSO does not use crossover and its mutation is defined through a vector addition. However, PSO differs from DE and most other EC dialects in that every candidate solution $\overline{x} \in \mathbb{R}^n$ carries its own perturbation vector $\overline{p} \in \mathbb{R}^n$. Technically, this makes them quite similar to evolution strategies that use the mutation step sizes in the perturbation vector parts, cf. Sect. 4.4.2 and Sect. 6.2. However, the PSO mindset and terminology is based on a spatial metaphor of particles with a location and velocity, rather than a biological one of individuals with a genotype and mutation.

To simplify the explanation and to emphasise the similarities to other evolutionary algorithms, we present PSOs in two steps. First we give a description that captures the essence of the system using the notion of perturbation vectors $\overline{p}$. Second, we provide the technical details in terms of vectors for velocity $\overline{v}$ and a personal best $\overline{b}$ in line with the PSO literature.

On a conceptual level every population member in a PSO can be considered as a pair $\langle \overline{x}, \overline{p} \rangle$, where $\overline{x} \in \mathbb{R}^n$ is a candidate solution vector and $\overline{p} \in \mathbb{R}^n$ is a perturbation vector that determines how the solution vector is changed to produce a new one. The main idea is that a new pair $\langle \overline{x}', \overline{p}' \rangle$ is produced from $\langle \overline{x}, \overline{p} \rangle$ by first calculating a new perturbation vector $\overline{p}'$ (using $\overline{p}$ and some additional information) and adding this to $\overline{x}$. That is,

$$\overline{x}' = \overline{x} + \overline{p}'$$

The core of the PSO perspective is to consider a population member as a point in space with a position and a velocity and use the latter to determine a new position (and a new velocity). Thus, looking under the hood of a PSO we find that a perturbation vector is a velocity vector $\overline{v}$ and a new velocity vector $\overline{v'}$ is defined as the weighted sum of three components: $\overline{v}$ and two vector differences. The first points from the current position $\overline{x}$ to the best position $\overline{y}$ the given population member ever had in the past, and the second points from $\overline{x}$ to the best position $\overline{z}$ the whole population ever had. Formally, we have

$$\overline{v}' = w \cdot \overline{v} + \phi_1 U_1 \cdot (\overline{y} - \overline{x}) + \phi_2 U_2 \cdot (\overline{z} - \overline{x})$$

where w and ϕ_i are the weights (w is called the inertia, ϕ_1 is the learning rate for the personal influence and ϕ_2 is the learning rate for the social influence), while U_1 and U_2 are randomizer matrices that multiply every coordinate of $\overline{y} - \overline{x}$ and $\overline{z} - \overline{x}$ by a number drawn from the uniform distribution.

It is worth noting that this mechanism requires some additional book keeping. In particular, the personal best $\overline{y}$ and the global best $\overline{z}$ must be kept in memory. This requires a unique identifier for population members that keeps the 'identity' of the given individuals and allows it to maintain the personal memory. To this end, PSO populations are lists, rather than (multi)sets, allowing references to the i-th individual. Similarly to DE, the order of individuals in such a population is not related to their fitness values. Furthermore, the perturbation vector $\overline{p}_i \in \mathbb{R}^n$ for any given $\overline{x}_i \in \mathbb{R}^n$ is not stored directly, as the notation $\langle \overline{x}, \overline{p} \rangle$ in the previous paragraph would indicate, but indirectly by the velocity vector $\overline{v}_i$ and the personal best $\overline{b}_i$ of the i-th population member. Thus, technically, the i-th individual is a triple $\langle \overline{x}_i, \overline{v}_i, \overline{b}_i \rangle$, where $\overline{x}_i$ is the solution vector (perceived as a position), $\overline{v}_i$ is its velocity vector, and $\overline{b}_i$ is its personal best. During an evolutionary cycle each triple $\langle \overline{x}_i, \overline{v}_i, \overline{b}_i \rangle$ is replaced by the mutant triple $\langle \overline{x}'_i, \overline{v}'_i, \overline{b}'_i \rangle$ using the following formulas

$$\overline{x}'_i = \overline{x} + \overline{v}'_i$$
$$\overline{v}'_i = w \cdot \overline{v}_i + \phi_1 U_1 \cdot (\overline{b}_i - \overline{x}_i) + \phi_2 U_2 \cdot (\overline{c} - \overline{x}_i)$$

where $\overline{c}$ denotes the population's global best (champion) and

$$\overline{b}'_i = \begin{cases} \overline{x}'_i & \text{if } f(\overline{x}'_i) < f(\overline{b}_i) \\ \overline{b}_i & \text{otherwise} \end{cases}$$

The rest of the basic PSO algorithm is actually quite simple, since parent selection and survivor selection are trivial. An overview is given in Table 6.7.

6.8 Estimation of Distribution Algorithms

Estimation of distribution algorithms (EDA) are based on the idea of replacing the creation of offspring by 'standard' variation operators (recombination

Representation	Real-valued vectors
Recombination	None
Mutation	Adding velocity vector
Parent selection	Deterministic (each parent creates one offspring via mutation)
Survival selection	Generational (offspring replace parents)

Table 6.7. Sketch of particle swarm optimisation

and mutation) by a three-step process. First a 'graphical model' is chosen to represent the current state of the search in terms of the dependencies between variables (genes) describing a candidate solution. Next the parameters of this model are estimated from the current population to create a conditional probability distribution over the variables. Finally, offspring are created by sampling this distribution.

Probabilistic Graphical Models (PGMs) have been used as models of uncertainty in artificial intelligence systems since the early 1980s. In this approach models are considered to be graphs $G = (V, E)$, where each vertex $v \in V$ represents a single variable, and each directed edge $e \in E$ represents a dependency between two variables. Thus, for example, the presence of an edge $e = \{i, j\}$ denotes that the probability of obtaining a particular value for variable j depends on the value of variable i. Usually graphs are restricted to be acyclic to avoid difficulties with infinite loops.

The pseudocode in Figure 6.2 illustrates the way that offspring are created via the processes of **model selection**, **model fitting** and **model sampling**.

```
BEGIN
  INITIALISE population P^0 with μ random candidate solutions;
  set t = 0;
  REPEAT UNTIL ( TERMINATION CONDITION is satisfied ) DO
    EVALUATE each candidate in P^t;
    SELECT subpopulation P_s^t to be used in the modeling steps;
    MODEL SELECTION creates graph G by dependencies in P_s^t;
    MODEL FITTING creates Bayesian Network BN with G and P_s^t;
    MODEL SAMPLING produces a set Sample(BN) of μ candidates;
    set t = t + 1;
    P^t = Sample(BN);
  OD
END
```

Fig. 6.2. Pseudocode for generic estimation of distribution algorithm

In principle, any standard selection method may be used to select P_s^t, but it is normal practice to use truncation selection and to take the fittest subset of the current population as the basis for the subsequent modelling process. Survivor selection in EDAs typically uses a generational model: newly generated offspring replaces the old population.

Model selection is the critical element in using a PGM approach to data modelling (in our case modelling a population in an EDA). In essence, it amounts to finding the appropriate structure to capture the conditional (in)dependencies between variables. In the context of genetics, and also of evolutionary computation, this process is also known as the "Linkage Learning" problem. Historically the pattern of research in this area has progressed via permitting increasing complexity for the types of structural dependencies examined. The earliest univariate algorithms such as PBIL [35] assumed variables behaved independently. The second generation of EDAs such as MIMIC [59] and BMDA [337] selected structures from the possible pairwise interactions, and current methods such as BOA [336] select structures from the set of all trees of some prespecified maximum size. Of course the number of possible combinations of variables expands extremely rapidly, meaning that some form of search is needed to identify good structural models. Broadly speaking there are two approaches to this. The first is direct estimation that has the reputation of being complex and in most cases impractical. More widely used is the "score + search" approach, which is effectively heuristic search in the space of possible graphical models. This relies heavily on the use of metrics which reflect how well a model, and hence the induced Bayesian Network, captures the underlying structure of the data. A variety of metrics have been proposed in the literature, mostly based on the Kullback–Liebler Divergence metric which is closely related to entropy. A full description of these different quality metrics is beyond the scope of this book.

The process of model-fitting may be characterised as per the pseudocode in Figure 6.3, where we use the notation $P(x, i, c)$ to denote the probability of obtaining allele value i for variable x given the set of conditions c. For the unconditional case we will use $P(x, i, -)$. It should be noted that this code is intended to illustrate the general concept, and that much more efficient implementations exist. For discrete data these parameters form a Bayesian Network, and for continuous data it is common to use a mixture of Gaussian models. Both of these processes are relatively straightforward.

The process of model sampling follows a similar pattern, but in this case within each subgraph we draw random variables to select the parent–node alleles, and then to select allele values for the other nodes using the probabilities from the appropriate partition.

Most of the discussion above has tacitly assumed discrete combinatorial problems and the models fitted have been Bayesian networks. For continuous variable problems a slightly different approach is needed to measure and describe probabilities, which is typically based on normal (Gaussian) distributions. There have been a number of developments, such as the **Iterated Den-**

```
BEGIN
  /*  Let P(x,i,c) denote the probability of generating */
  /*  allele value i for variable x given conditions c */
  /*  Let D denote the set of selected parents */
  /*  Let G denote the model selected */

  FOR EACH unconnected subgraph g ⊂ G DO
    FOR EACH node x ∈ g with no parents DO
      FOR EACH possible allele value i for variable x DO
        set P(x,i,−) = Frequency_In_Subpop(x,i,D);
      OD
    OD
    FOR EACH child node x ∈ g DO
      Partition D according to allele values in x's parents;
      FOR EACH partition c DO
        set P(c) = Sizeof(c)/Sizeof(D);
        FOR EACH possible allele value i for variable x DO
          set P(x,i,c) = Frequency_In_Subpop(x,i,c)/P(c);
        OD
      OD
    OD
  OD
END
```

Fig. 6.3. Pseudocode for generic model fitting

sity Estimation Algorithm [64], and also continuous versions for EDAs. Depending on the complexity of the models permitted, the result is either a univariate or multivariate normal distribution. These are very similar to the correlation matrix in evolution strategies, and a comparison of these approaches may be found in [206].

For exercises and recommended reading for this chapter, please visit www.evolutionarycomputation.org.

Part II

Methodological Issues

7

Parameters and Parameter Tuning

Chapter 3 presented an algorithmic framework that forms the common basis for all evolutionary algorithms. A decision to use an evolutionary algorithm implies that the user adopts the main design decisions behind this framework. Thus, the main algorithm setup follows automatically: the algorithm is based on a population of candidate solutions that is manipulated by selection, recombination, and mutation operators. To obtain a concrete, executable EA, the user only needs to specify a few details. In this chapter we have a closer look at these details, named **parameters**. We discuss the notion of EA parameters and explain why the task of designing an evolutionary algorithm can be seen as the problem of finding appropriate parameter values. Furthermore, we elaborate on the problem of tuning EA parameters and provide an overview of different algorithms that can tune EAs with limited user effort.

7.1 Evolutionary Algorithm Parameters

For detailed discussion of the notion of EA parameters, let us consider a simple GA. As explained in Section 6.1, this is a well-established algorithm with a few degrees of freedom, including the parameters `crossoveroperator`, `crossoverrate`, and `populationsize` (and some others we do not need for the present discussion). To obtain a fully specified, executable version we must provide specific values for these parameters, for instance, setting the parameter `crossoveroperator` to `onepoint`, the parameter `crossoverrate` to 0.5, and the parameter `populationsize` to 100. In principle, we need not distinguish different types of parameters, but intuitively there is a difference between deciding on the crossover operator to be used and choosing a value for the related crossover rate. This difference can be formalized if we distinguish parameters by their domains. The parameter `crossoveroperator` has a finite domain with no sensible distance metric or ordering, e.g., {`onepoint`, `uniform`, `averaging`}, whereas the domain of the parameter $p_c \in [0, 1]$ is a subset of the real numbers $\mathbb{R}$ with the natural structure for real

Table 7.1. Pairs of terms used in the literature to distinguish two types of parameters (variables).

Parameter with an unordered domain	Parameter with an ordered domain
qualitative	quantitative
symbolic	numeric
categorical	numerical
structural	behavioral
component	parameter
nominal	ordinal
categorical	ordered

numbers. This difference is essential for searchability. For parameters with a domain that has a distance metric, or is at least partially ordered, one can use heuristic search and optimization methods to find optimal values. For the other type of parameters this is not possible because the domain has no exploitable structure. The only option in this case is sampling.

The difference between two types of parameters has already been noted in evolutionary computing, but various authors use various naming conventions as shown in Table 7.1. Table 7.2 shows an EA-specific illustration with commonly used parameters in both categories. Throughout this book we use the terms **symbolic parameter** and **numeric parameter**. For both types of parameters the elements of the parameter's domain are called **parameter values** and we instantiate a parameter by allocating a value to it.

It is important to note that, depending on particular design choices, one might obtain different numbers of parameters for an EA. For instance, instantiating the symbolic parameter `parentselection` by `tournament` implies a new numeric parameter `tournamentsize`. However, choosing `roulettewheel` does not add any parameters. This example also shows that there can be a hierarchy among parameters. Namely, symbolic parameters may have numeric parameters 'under them'.

7.2 EAs and EA Instances

The distinction between symbolic and numeric parameters naturally supports a distinction between EAs and EA instances. To be specific, we can consider symbolic parameters as high-level ones that define the essence of an evolutionary algorithm, and look at numeric parameters as low-level ones that define a specific variant of this EA. Following this naming convention, an **evolutionary algorithm** is a partially specified algorithm fitting the framework introduced in Chapter 3, where the values to instantiate symbolic parameters are defined, but the numeric parameters are not. Hence, we consider two EAs to be different if they differ in one or more of their symbolic parameters, for instance, if they use different mutation operators. If the values are specified for all parameters, including the numeric ones then we obtain an **evolutionary algorithm instance.** If two EA instances differ only in some of the values of their numeric parameters (e.g., the mutation rate and the tournament size), then we consider them as two variants of the same EA. Table 7.2 illustrates this matter by showing three EA instances belonging to just two EAs.

	A_1	A_2	A_3
symbolic parameters			
`representation`	bitstring	bitstring	real-valued
`recombination`	1-point	1-point	averaging
`mutation`	bit-flip	bit-flip	Gaussian $N(0,\sigma)$
`parentselection`	tournament	tournament	uniform random
`survivorselection`	generational	generational	(μ, λ)
numeric parameters			
p_m	0.01	0.1	0.05
σ	n.a.	n.a.	0.1
p_c	0.5	0.7	0.7
μ	100	100	10
λ	equal μ	equal μ	70
κ	2	4	n.a.

Table 7.2. Three EA instances specified by the symbolic parameters `representation`, `recombination`, `mutation`, `parentselection`, `survivorselection`, and the numeric parameters `mutationrate` (p_m), `mutationstepsize` (σ), `crossoverrate` (p_c), `populationsize` (μ), `offspringsize` (λ), and `tournamentsize` (κ). In our terminology, the instances in columns A_1 and A_2 are just variants of the same EA. The EA instance in column A_3 is an example of a different EA, because it has different symbolic parameter values.

This terminology enables precise formulations and enforces care with phrasing. Observe that the distinction between EAs and EA instances is similar to distinguishing between problems and problem instances. If rigorous terminology is required then the right phrasing is "to apply an EA instance to a problem instance". However, such rigour is not always needed, and formally inaccurate but understandable phrases like "to apply an EA to a problem" are acceptable if they cannot lead to confusion.

7.3 Designing Evolutionary Algorithms

In the broad sense, algorithm design includes all the decisions needed to specify an algorithm (instance) to solve a given problem (instance). The principal challenge for evolutionary algorithm designers is that the design details, i.e., parameter values, have such a large influence on the performance of the algorithm. Hence, the design of algorithms in general, and EAs in particular, is an optimization problem itself.

To understand this issue, we distinguish three layers: application, algorithm, and design, as shown in Figure 7.1. The whole scheme can be divided into two optimization problems that we refer to as problem solving (the lower part) and algorithm design (the upper part). The lower part consists of an EA instance at the algorithm layer that is trying to find an optimal solution for the given problem instance at the application layer. The upper part contains a design method — the intuition and heuristics of a human user or an automated design strategy — that is trying to find optimal parameter values for the given EA at the algorithm layer. The quality of a given parameter vector is based on the performance of the EA instance using these values.

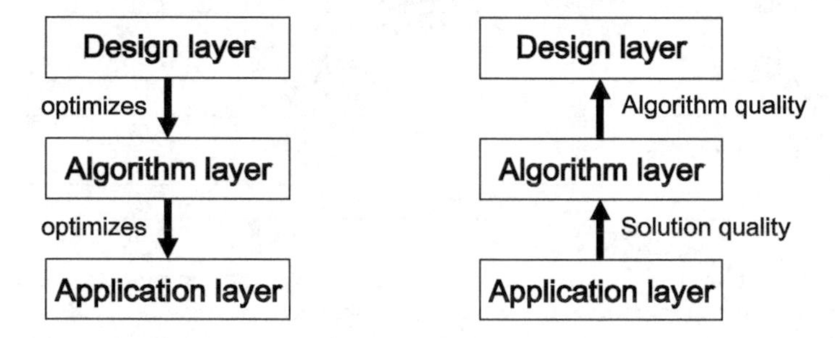

Fig. 7.1. Control flow (left) and information flow (right) through the three layers in the hierarchy of algorithm design. Left: the entity on a given layer optimises the entity on the layer below. Right: the entity on a given layer provides information to the entity on the layer above.

To avoid confusion we use distinct terms to designate the quality function of these optimization problems. In keeping with the usual EC terminology we use the term **fitness** at the application layer, and the term **utility** at the algorithm layer. In the same spirit, we use the term **evaluation** only in relation to fitness, cf. fitness evaluation, and **testing** in relation to utility. With this nomenclature, the problem to be solved by the algorithm designer can be seen as an optimization problem in the space of parameter vectors given some utility function. Solutions of the EA design problem are therefore EA parameter vectors with maximum utility. Table 7.3 provides a quick overview of the resulting vocabulary.

Now we can define the **utility landscape** as an abstract landscape where the locations are the parameter vectors of an EA and the height reflects utility. It is obvious that fitness landscapes, commonly used in EC, have a lot in common with utility landscapes as introduced here. However, despite the obvious analogies, there are some differences we want to note. First of all, fitness values are typically deterministic for most problems. However, utility values are always stochastic, because they reflect the performance of an EA which is a stochastic search method. This implies that the maximum utility needs to be defined in some statistical sense. Consequently, comparing EA parameter vectors can be difficult if the underlying data produced by different EA runs shows a big variance. Second, the notion of fitness is usually strongly related to the objective function of the problem in the application layer, and differences between suitable fitness functions mostly concern arithmetic details. In contrast, the notion of utility depends on the performance metrics used, which reflect the preferences of the user and the context in which the EA is being used. For example, solving a single problem instance just once or repeatedly solving instances of the same problem type represent two very different use cases with different implications for the optimal EA (instance). This issue is dealt with in more detail in Chap. 9.

7.4 The Tuning Problem

To recap, producing an executable EA instance requires specifying values for its parameters. These values determine whether it will find an optimal solution, and whether it will do so efficiently. Parameter tuning is a commonly practised approach to algorithm design, where the values for the parameters are established *before* the run of the algorithm and they remain fixed during the run.

The common way to solve the tuning problem is based on conventions ("mutation rate should be low"), ad hoc choices ("why not use population size 100") and limited experimentation with different values, for example, considering four parameters and five values for each of them. The drawbacks to the first two are obvious. The problems with the experimentation-based approach are traditionally summarised as follows:

- Parameter effects interact (for example, diversity can be created by recombination or mutation), hence they cannot be optimised one by one.
- Trying all different combinations systematically is extremely time consuming. Testing five different values for four parameters leads to $5^4 = 625$ different setups. Performing 100 independent EA runs with each setup implies 62,500 runs with the EA before we can even start the 'real' run.

	Problem solving	**Algorithm design**
Method at work	evolutionary algorithm	design procedure
Search space	solution vectors	parameter vectors
Quality	fitness	utility
Assessment	evaluation	testing

Table 7.3. Vocabulary for problem solving and algorithm design.

- For a numerical parameter, the best parameter values may not even be among the ones we selected for testing. This is because an optimal parameter value could very well lie between the points in the grid we were testing. Increasing the resolution by increasing the number of grid points increases the number of runs exponentially.

This picture becomes even more discouraging if one is after a generally good setup that would perform well on a range of problems. During the history of EAs considerable effort has been spent on finding parameter values that worked well for a number of test problems. A well-known early example is De Jong's thesis [102], which determined recommended values for the probabilities of single-point crossover and bit mutation on what is now called the De Jong test suite of five functions. The contemporary view of EAs, however, acknowledges that specific problem (instances) may require specific EA setups for satisfactory performance [26]. Thus, the scope of 'optimal' parameter settings is necessarily narrow. There are also theoretical arguments that *any* quest for generally good parameter settings is lost a priori, cf. the discussion of the No Free Lunch theorem [467] in Chap. 16.

During the first decades of the evolutionary computing history the tuning issue was largely neglected. Scientific publications did not provide any justification for the parameter values used, nor did they describe the effort spent at arriving to these parameter values. Consequently, it was impossible to tell if the reported EA performance was exceptional (only achievable through intensive tuning) or trivial (obtained by just a bit of tweaking with the parameter values). There was not much research into developing decent tuning procedures either. The idea of optimising GA parameters by a meta-GA was suggested in 1986 [202], but algorithmic approaches to parameter tuning did not receive significant attention for a long time. This situation changed around 2005, when several good tuning algorithms were proposed within a short time interval, such as SPO [42, 41, 43], (iterative) F-race [55, 33], and REVAC [313, 314] and the idea of the meta-GA was also revived [471]. In the last ten years parameter tuning has reached a mature stage: the most important issues are well understood and there are various tuning algorithms available [145, 125]. There are experimental comparisons between some of these methods [378]. However, large-scale adoption by EC researchers and practitioners is still ahead of us.

The basis of the modern approach to parameter tuning is to consider the design of an evolutionary algorithm as a search problem and that of a tuning method as a search algorithm. To this end, it is important that a search algorithm generates a lot of data. In our case, these data concern parameter vectors and their utility values. If one is only interested in an optimal EA configuration then such data are not relevant – finding a good parameter vector is enough. However, these data can be used to reveal information about the given evolutionary algorithm, for instance on its robustness, distribution of solution quality, sensitivity, etc. Thus, adopting the terminology of Hooker

[221], parameter tuning can then be used for competitive as well as scientific testing:

- to configure an EA by choosing parameter values that optimise its performance, and
- to analyse an EA by studying how its performance depends on its parameter values and/or the problems it is applied to.

In both cases, the solutions of a tuning problem depend on the problem(s) to be solved, the EA used, and the utility function that defines how we measure algorithm quality. Adding the tuner to the equation, we obtain Fig. 7.2 which illustrates the generic scheme of parameter tuning in graphical form.

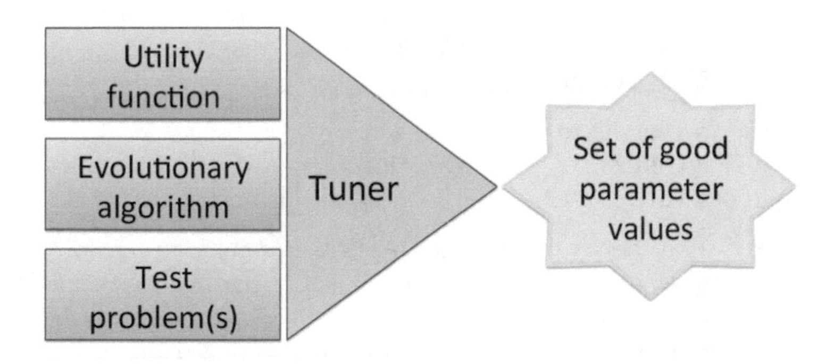

Fig. 7.2. Generic scheme of parameter tuning showing how good parameter values depend on four factors: the problem instance(s) to be solved, the EA used, the utility function, and the tuner itself.

7.5 Algorithm Quality: Performance and Robustness

In general, there are two basic performance measures for EAs, one regarding solution quality and one regarding algorithm speed. Most, if not all, metrics used in EC are based on variations and combinations of these two. Solution quality can be naturally expressed by the fitness function. As for algorithm speed, time or search effort needs to be measured. This can be done by, for instance, the number of fitness evaluations, CPU time, wall-clock time, etc. In [124] and in Chap. 9 of this book we discuss the pros and cons of various time measures. Here we do not go into this issue, we just assume that one of them has been chosen. Then there are three basic combinations of solution quality and computing time that can be used to define algorithm performance in a single run: fix time and measure quality; fix quality and measure time; or fix both and measure completion. For instance:

- Given a maximum running time (computational effort), performance is defined as the best fitness at termination.
- Given a minimum fitness level, performance is defined as the running time (computational effort) needed to reach it.
- Given a maximum running time (computational effort) and a minimum fitness level, performance is defined through the Boolean notion of success: a run is deemed successful if the given fitness is reached within the given time.

Because of the stochastic nature of EAs, a good estimation of performance requires multiple runs on the same problem with the same parameter values and some statistical aggregation of the measures defined for single runs. Doing this for the three measures above gives us the performance metrics commonly used in evolutionary computing:

- MBF (**mean best fitness**),
- AES (**average number of evaluations to a solution**),
- SR (**success rate**).

In Chapter 9 we discuss performance measures in more detail. For this discussion we merely note that it is the choice of performance metrics that determines the utility landscape, and therefore which parameter vector is best. It was recently shown that tuning for different performance measures can yield parameter values that differ in orders of magnitude [376]. This demonstrates why any claim about good parameter values in general, without a reference to the performance measure, should be taken with a pinch of salt.

Regarding **robustness**, the first thing to be noted is that there are different interpretations of this notion in the literature. The existing (informal) definitions do agree that robustness is related to the variance of an algorithm's performance across some dimension, but they differ in what this dimension is. There are indeed more options here, given the fact that the performance of an EA (instance) depends on (1) the problem instance it is solving, (2) the parameter vector it uses, and (3) effects from the random number generator. Therefore, the variance of performance can be considered along three different dimensions: parameter values, problem instances, and random seeds, leading to three different types of robustness.

The first type of robustness is encountered if we are tuning an evolutionary algorithm A on a test suite consisting of many problem instances or test functions. The result of the tuning process is a parameter vector $\bar{p}$ and the corresponding EA instance $A(\bar{p})$ that exhibits good performance over the whole test suite. Note that in this case robustness is defined for EA instances, not EAs. For the historically inclined readers, the famous figures in the classic books of Goldberg [189, page 6], and Michalewicz, [296, page 292], refer to this kind of robustness, with respect to a range of problems.

Another popular interpretation of algorithm robustness is related to performance variations caused by different parameter values. This notion of robustness is defined for EAs. Of course, it is again the EA instance $A(\bar{p})$ whose performance forms the basic measurement, but here we aggregate over parameter vectors. Using such a definition, it is EAs (specified by a particular configuration of the symbolic parameters) that can be compared by their robustness. Finding robust EAs in this sense requires a search through the symbolic parameters.

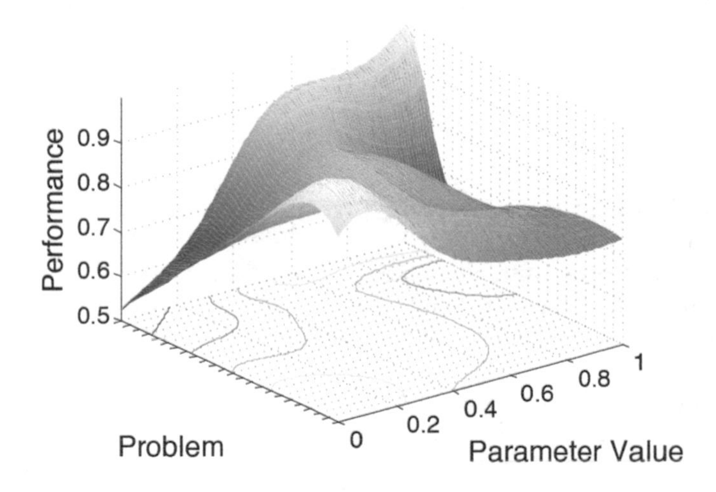

Fig. 7.3. Illustration of the grand utility landscape showing the performance (z) of EA instances belonging to a given parameter vector (x) on a given problem instance (y). Note: The 'cloud' of repeated runs is not shown.

Figure 7.3 shows the grand utility landscape based on all possible combinations of EA parameters and problem instances. For the sake of this illustration we only take a single parameter into account. Thus, we obtain a 3D landscape with one axis, x, representing the values of the parameter and another axis, y, representing the problem instances investigated. (In the general case of n parameters, we have $n+1$ axes here.) The third dimension, z, shows the performance of the EA instance belonging to a given parameter vector on a given problem instance. It should be noted that for stochastic algorithms, such as EAs, this landscape is blurry if the repetitions with different random seeds are also taken into account. That is, rather than one z-value for a pair $\langle x, y \rangle$, we have one z for every run, for repeated runs we get a 'cloud'.

Although this 3D landscape gives the best complete overview of EA performance and robustness, lower-dimensional hyperplanes are also interesting and can be more revealing. The left-hand side of Figure 7.4 shows 2D slices corresponding to specific parameter vectors, i.e., EA instances. Such a slice shows how the performance of an EA instance varies over the range of problem

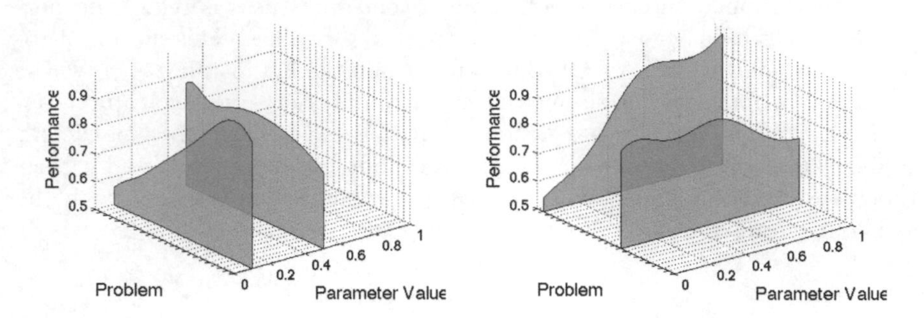

Fig. 7.4. Illustration of parameter-wise slices (left) and problem-wise slices (right) of the grand utility landscape shown in Figure 7.3. (The 'cloud' of repeated runs is not shown.) See the text for explanation and interpretation.

instances. This provides information on robustness with regard to changes in problem specification. Such data are often reported in the literature, often in the form of tables containing the experimental results of one or more EA instances on a predefined test suite, such as the five De Jong functions, the 25 functions of the CEC 2005 contest [420], and the GECCO 2010 test suite [13].

The right-hand side of Figure 7.4 shows 2D slices corresponding to specific problem instances. Each slice shows how the performance of the given EA depends on the parameter values it uses, i.e., its robustness to changes in parameter values. In evolutionary computing such data is hardly ever published. This is a straightforward consequence of the current practice, where parameter values are mostly selected by conventions, ad hoc choices, and very limited experimental comparisons. In other words, usually such data is not even produced, let alone stored and presented. By the increased adoption of tuning algorithms this practice could change, and knowledge about EA parameterization could be collected and disseminated.

7.6 Tuning Methods

In essence, all tuning algorithms work by the GENERATE-and-TEST principle, i.e., by generating parameter vectors and testing them to establish their utility. Considering the GENERATE step, tuners can be then divided into two main categories: *non-iterative* and *iterative* tuners. All non-iterative tuners execute the GENERATE step only once, during initialization, thus creating a fixed set of vectors. Each of those vectors is then tested during the TEST phase to find the best vector in the given set. Hence, one could say that non-iterative tuners follow the INITIALIZE-and-TEST template. Initialization can be done by random sampling, generating a systematic grid in the parameter space, or some space filling set of vectors. Examples of such methods are Latin-Square citemyers2001empirical-model and Taguchi Orthogonal Arrays [424]. Perhaps the best-known method in this category is the frequently used parameter 'optimisation' through a systematic comparison of a few combinations of parameter values, e.g., four mutation rates, four crossover rates, two values for tournament size, and four values for population size. In contrast, iterative tuners do not fix the set of vectors during initialization, but start with a small initial set and create new vectors iteratively during execution. Common examples of such methods are meta-EAs and iterative sampling methods.

Considering the TEST step, we can again distinguish two types of tuners: *single-stage* and *multi-stage procedures*. In both cases the tuners perform a number of tests (i.e., EA runs with the given parameter values) for a reliable estimate of utility. This is necessary because of the stochastic nature of EAs. The difference between the two types is that single-stage procedures perform the same number of tests for each given vector, while multi-stage procedures use a more sophisticated strategy. In general, they augment the TEST step by adding a SELECT step, where only promising vectors are selected for further testing, deliberately ignoring those with a low performance. The best-known method to this end is racing [283].

A further useful distinction between tuning methods can be made by their use of meta-models of the utility landscape. From this perspective tuners can be divided into two major classes: model-free and model-based approaches [226]. Meta-EAs, ParamILS [227], and F-Race [56] belong to the first category. They are 'simply' optimising the given utility landscape, trying to find parameter vectors that maximise the performance of the EA to be tuned. SPO, REVAC, and Bonesa [377] do more than that: during the tuning process they create a model that estimates the performance of an EA for any given parameter vector. In other words, they use meta models or surrogate models [153, 235] that have two advantages. First, meta models reduce the number of expensive utility tests by replacing some of the real tests by model estimates that can be calculated very quickly. Second, they capture information about parameters and their utility for algorithm analysis.

In summary, there exist good tuners that are able to find good parameter values for EAs, and in principle they can tune any heuristic search method with parameters. There are differences between them in terms of efficiency (time needed for tuning), solution quality (performance of the optimised EA), and ease of use (e.g., the number of their own parameters). However, as of 2013, a solid experimental comparison of such tuners is not available. For more details we refer to two recent survey papers [145, 125].

For exercises and recommended reading for this chapter, please visit `www.evolutionarycomputation.org`.

8

Parameter Control

The issue of setting the values of evolutionary algorithm parameters *before* running an EA was treated in the previous chapter. In this chapter we discuss how to do this *during* a run of an EA, in other words, we elaborate on controlling EA parameters on-the-fly. This has the potential of adjusting the algorithm to the problem while solving the problem. We provide a classification of different approaches based on a number of complementary features and present examples of control mechanisms for every major EA component. Thus we hope to both clarify the points we wish to raise and also to give the reader a feel for some of the many possibilities available for controlling different parameters.

8.1 Introduction

In the previous chapter we argued that parameter tuning can greatly increase the performance of EAs. However, the tuning approach has an inherent drawback: parameter values are specified before the run of the EA and these values remain fixed during the run. But a run of an EA is an intrinsically dynamic, adaptive process. The use of rigid parameters that do not change their values is thus in contrast to this spirit. Additionally, it is intuitively obvious, and has been empirically and theoretically demonstrated on many occasions, that different values of parameters might be optimal at different stages of the evolutionary process. For instance, large mutation steps can be good in the early generations, helping the exploration of the search space, and small mutation steps might be needed in the late generations for fine-tuning candidate solutions. This implies that the use of static parameters itself can lead to inferior algorithm performance.

A straightforward way to overcome the limitations of static parameters is by replacing a parameter p by a function $p(t)$, where t is the generation counter (or any other measure of elapsed time). However, as discussed in Chap. 7, the problem of finding optimal static parameters for a particular problem is already hard. Designing optimal dynamic parameters (that is, functions for $p(t)$) may be even more difficult. Another drawback to this approach is that the parameter value $p(t)$ changes are caused by a 'blind' deterministic rule triggered by the progress of time t, unaware of the current state of the search. A well-known instance of this problem occurs in simulated annealing (Sect. 8.4.5) where a so-called cooling schedule has to be set before the execution of the algorithm.

Mechanisms for modifying parameters during a run in an 'informed' way were realised quite early in EC history. For instance, evolution strategies changed mutation parameters on-the-fly by Rechenberg's 1/5 success rule [352] using information about the ratio of successful mutations. Davis experimented with changing the crossover rate in GAs based on the progress realised by particular crossover operators [97]. The common feature of such methods is the presence of a human-designed feedback mechanism that utilises actual information about the search to determine new parameter values.

A third approach is based on the observation that finding good parameter values for an EA is a poorly structured, ill-defined, complex problem. This is exactly the kind of problem on which EAs are often considered to perform better than other methods. It is thus a natural idea to use an EA for tuning an EA to a particular problem. This could be done using a meta-EA or by using only one EA that tunes itself to a given problem *while* solving that problem. Self-adaptation, as introduced in evolution strategies for varying the mutation parameters, falls within this category. In the next section we discuss various options for changing parameters, illustrated by an example.

8.2 Examples of Changing Parameters

Consider a numerical optimisation problem of minimising

$$f(\overline{x}) = f(x_1, \ldots, x_n),$$

subject to some inequality and equality constraints

$$g_i(\overline{x}) \leq 0, i = 1, \ldots, q,$$

and

$$h_j(\overline{x}) = 0, j = q+1, \ldots, m,$$

where the domains of the variables are given by lower and upper bounds $l_i \leq x_i \leq u_i$ for $1 \leq i \leq n$. For such a problem we might design an EA based on a floating-point representation, where each individual $\overline{x}$ in the population is represented as a vector of floating-point numbers $\overline{x} = \langle x_1, \ldots, x_n \rangle$.

8.2.1 Changing the Mutation Step Size

Let us assume that offspring are produced by arithmetic crossover followed by Gaussian mutation that replaces components of the vector $\overline{x}$ by

$$x'_i = x_i + N(0, \sigma).$$

To adjust σ over time we use a function $\sigma(t)$ defined by some heuristic rule and a given measure of time t. For example, the mutation step size may be defined by the current generation number t as:

$$\sigma(t) = 1 - 0.9 \cdot \frac{t}{T},$$

where t varies from 0 to T, the maximum generation number. Here, the mutation step size $\sigma(t)$, which is used for all for vectors in the population and for all variables of each vector, decreases slowly from 1 at the beginning of the run ($t = 0$) to 0.1 as the number of generations t approaches T. Such decreases may assist the fine-tuning capabilities of the algorithm. In this approach, the value of the given parameter changes according to a fully deterministic scheme. The user thus has full control of the parameter, and its value at a given time t is completely determined and predictable.

Second, it is possible to incorporate feedback from the search process, still using the same σ for all vectors in the population and for all variables of each vector. For example, Rechenberg's 1/5 success rule [352] states that the ratio of successful mutations to all mutations should be 1/5. Hence if the ratio is greater than 1/5 the step size should be increased, and if the ratio is less than 1/5 it should be decreased. The rule is executed at periodic intervals, for instance, after k iterations each σ is reset by

$$\sigma' = \begin{cases} \sigma / c & \text{if} \quad p_s > 1/5, \\ \sigma \cdot c & \text{if} \quad p_s < 1/5, \\ \sigma & \text{if} \quad p_s = 1/5, \end{cases}$$

where p_s is the relative frequency of successful mutations, measured over a number of trials, and the parameter c should be $0.817 \leq c \leq 1$ [372]. Using this mechanism, changes in the parameter values are now based on feedback from the search. The influence of the user is much less direct here than in the deterministic scheme above. Of course, the mechanism that embodies the link between the search process and parameter values is still a heuristic rule indicating how the changes should be made, but the values of $\sigma(t)$ are not deterministic.

Third, we can assign an individual mutation step size to each solution and make these co-evolve with the values encoding the candidate solutions. To this end we extend the representation of individuals to length $n + 1$ as $\langle x_1, \ldots, x_n, \sigma \rangle$ and apply some variation operators (e.g., Gaussian mutation and arithmetic crossover) to the values of x_i as well as to the σ value of an individual. In this way, not only the solution vector values (x_i) but also the mutation step size of an individual undergo evolution. A solution introduced in Sect. 4.4.2 is:

$$\sigma' = \sigma \cdot e^{\tau \cdot N(0,1)}, \tag{8.1}$$

$$x'_i = x_i + \sigma' \cdot N_i(0,1). \tag{8.2}$$

Observe that within this self-adaptive scheme the heuristic character of the mechanism resetting the parameter values is eliminated, and a certain value of σ acts on all values of a single individual.

Finally, we can use a separate σ_i for each x_i, extend the representation to

$$\langle x_1, \ldots, x_n, \sigma_1, \ldots, \sigma_n \rangle,$$

and use the mutation mechanism described in Eq. (4.4). The resulting system is the same as the previous one, except the granularity, here we are co-evolving n parameters of the EA instead of 1.

8.2.2 Changing the Penalty Coefficients

In this section we illustrate that the evaluation function (and consequently the fitness function) can also be parameterised and varied over time. While this is a less common option than tuning variation operators, it can provide a useful mechanism for increasing the performance of an EA.

When dealing with constrained optimisation problems, penalty functions are often used (see Chap. 13 for more details). A common technique is the method of static penalties [302], which requires penalty parameters within the evaluation function as follows:

$$eval(\overline{x}) = f(\overline{x}) + W \cdot penalty(\overline{x}),$$

where f is the objective function, $penalty(\overline{x})$ is zero if no violation occurs and is positive[1] otherwise, and W is a user-defined weight prescribing how severely constraint violations are weighted. For instance, a set of functions f_j ($1 \le j \le m$) can be used to construct the penalty, where the function f_j measures the violation of the jth constraint:

$$f_j(\overline{x}) = \begin{cases} \max\{0, g_j(\overline{x})\} & \text{if} \quad 1 \le j \le q, \\ |h_j(\overline{x})| & \text{if} \quad q+1 \le j \le m. \end{cases} \tag{8.3}$$

To adjust the evaluation function over time, we can replace the static parameter W by a function $W(t)$. For example, the method in [237] uses

$$W(t) = (C \cdot t)^{\alpha},$$

[1] For minimisation problems.

where C and α are constants. Note that the penalty pressure grows with the evolution time provided $1 \leq C$ and $1 \leq \alpha$.

A second option is to utilise feedback from the search process. In one example, the method decreases the penalty component $W(t+1)$ for the generation $t+1$ if all best individuals in the last k generations were feasible, and increases penalties if all best individuals in the last k generations were infeasible. If there are some feasible and infeasible individuals as best individuals in the last k generations, $W(t+1)$ remains without change, cf. [45]. Technically, $W(t)$ is updated in every generation t in the following way:

$$W(t+1) = \begin{cases} (1/\beta_1) \cdot W(t) & \text{if } \overline{b}^i \in \mathcal{F} \quad\;\;\; \text{for all } t-k+1 \leq i \leq t, \\ \beta_2 \cdot W(t) & \text{if } \overline{b}^i \in \mathcal{S}-\mathcal{F} \text{ for all } t-k+1 \leq i \leq t, \\ W(t) & \text{otherwise.} \end{cases}$$

In this formula, $\mathcal{S}$ is the set of all search points (solutions), $\mathcal{F} \subseteq \mathcal{S}$ is a set of all *feasible* solutions, $\overline{b}^i$ denotes the best individual in terms of the function *eval* in generation i, $\beta_1, \beta_2 > 1$, and $\beta_1 \neq \beta_2$ (to avoid cycling).

Third, we could allow self-adaptation of the weight parameter, similarly to the mutation step sizes in the previous section. For example, it is possible to extend the representation of individuals into $\langle x_1, \ldots, x_n, W \rangle$, where W is the weight that undergoes the same mutation and recombination as any other variable x_i. Furthermore, we can introduce a separate penalty for each constraint as per Eq. (8.3). Hereby we obtain a vector of weights and can extend the representation to $\langle x_1, \ldots, x_n, w_1, \ldots, w_m \rangle$. Then define

$$eval(\overline{x}) = f(\overline{x}) + \sum_{j=1}^{m} w_j f_j(\overline{x}),$$

as the function to be minimised. Variation operators can then be applied to both the $\overline{x}$ and the $\overline{w}$ part of these chromosomes, realising a self-adaptation of the penalties, and thereby the fitness function.

It is important to note the crucial difference between self-adapting mutation step sizes and constraint weights. Even if the mutation step sizes are encoded in the chromosomes, the evaluation of a chromosome is *independent* from the actual σ values. That is,

$$eval(\langle \overline{x}, \overline{\sigma} \rangle) = f(\overline{x}),$$

for any chromosome $\langle \overline{x}, \overline{\sigma} \rangle$. In contrast, if constraint weights are encoded in the chromosomes, then we have

$$eval(\langle \overline{x}, \overline{w} \rangle) = f_{\overline{w}}(\overline{x}),$$

for any chromosome $\langle \overline{x}, \overline{w} \rangle$. This could enable the evolution to 'cheat' in the sense of making improvements by minimising the weights instead of optimising f and satisfying the constraints. Eiben et al. investigated this issue in [134] and found that using a specific tournament selection mechanism neatly solves this problem and enables the EA to solve constraints.

8.3 Classification of Control Techniques

In classifying parameter control techniques of an evolutionary algorithm, many aspects can be taken into account. For example:

1. *what* is changed (e.g., representation, evaluation function, operators, selection process, mutation rate, population size, and so on)
2. *how* the change is made (i.e., deterministic heuristic, feedback-based heuristic, or self-adaptive)
3. *the evidence* upon which the change is carried out (e.g., monitoring performance of operators, diversity of the population, and so on)
4. *the scope/level* of change (e.g., population-level, individual-level, and so forth)

In the following we discuss these items in more detail.

8.3.1 *What* Is Changed?

To classify parameter control techniques from the perspective of what component or parameter is changed, it is necessary to agree on a list of all major components of an evolutionary algorithm, which is a difficult task in itself. For that purpose, let us assume the following components of an EA:

- representation of individuals
- evaluation function
- variation operators and their probabilities
- selection operator (parent selection or mating selection)
- replacement operator (survival selection or environmental selection)
- population (size, topology, etc.)

Note that each component can be parameterised, and that the number of parameters is not clearly defined. Despite the somewhat arbitrary character of this list of components and of the list of parameters of each component, we will maintain the 'what-aspect' as one of the main classification features, since this allows us to locate where a specific mechanism has its effect.

8.3.2 *How* Are Changes Made?

As illustrated in Sect. 8.2, methods for changing the value of a parameter (i.e., the 'how-aspect') can be classified into one of three categories.

- **Deterministic parameter control**
  Here the value of a parameter is altered by some deterministic in rule predetermined (i.e., user-specified) manner without using any feedback from the search. Usually, a time-varying schedule is used, i.e., the rule is activated at specified intervals.

- **Adaptive parameter control**
  Here some form of feedback from the search serves as input to a mechanism that determines the change. Updating the parameter values may involve credit assignment, based on the quality of solutions discovered by different operators/parameters, so that the updating mechanism can distinguish between the merits of competing strategies. The important point to note here is that the updating mechanism used to control parameter values is externally supplied, rather than being part of the usual evolutionary cycle.
- **Self-adaptive parameter control**
  Here the evolution of evolution is used to implement the self-adaptation of parameters [257]. The parameters to be adapted are encoded into the chromosomes and undergo mutation and recombination. The better values of these lead to better individuals, which in turn are more likely to survive, produce offspring and hence propagate these better parameter values. This is an important distinction between adaptive and self-adaptive schemes: in the latter the mechanisms for the credit assignment and updating of different strategy parameters are entirely implicit, i.e., they are the selection and variation operators of the evolutionary cycle itself.

This terminology leads to the taxonomy illustrated in Fig. 8.1.

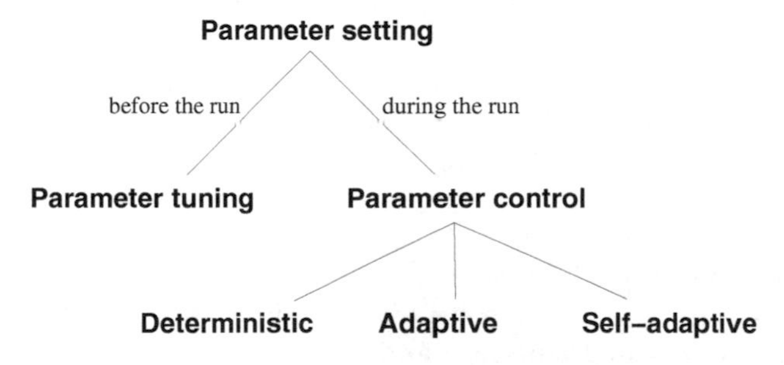

Fig. 8.1. Global taxonomy of parameter setting in EAs

Some authors have introduced a different terminology, cf. [8] or [410], but after the publication of [133] the one in Fig. 8.1 became the most widely accepted one. However, we acknowledge that the terminology proposed here is not perfect. For instance, the term “deterministic” control might not be the most appropriate, as it is not determinism that matters, but the fact that the parameter-altering mechanism is “uninformed”, i.e., takes no input related to the progress of the search process. For example, one might *randomly* change the mutation probability after every 100 generations, which is not a deterministic process. Also, the terms “adaptive” and “self-adaptive” could be replaced by the equally meaningful “explicitly adaptive” and “implicitly adaptive”, respectively. We have chosen to use “adaptive” and “self-adaptive” because of the widely accepted usage of the latter term.

8.3.3 What *Evidence* Informs the Change?

The third criterion for classification concerns the evidence used for determining the change of parameter value [382, 397]. Most commonly, the progress of the search is monitored by looking at the performance of operators, the diversity of the population, and so on, and the information gathered is used as feedback for adjusting the parameters. From this perspective, we can further distinguish between the following two cases:

- **Absolute evidence**
  We speak of absolute evidence when the rule to change the value of a parameter is applied when a predefined event occurs. For instance, one could increase the mutation rate when the population diversity drops under a given value, or resize the population based on estimates of schemata fitness and variance. As opposed to deterministic parameter control, where a rule fires by a deterministic trigger (e.g., time elapsed), here feedback from the search is used. Such mechanisms require that the user has a clear intuition about how to steer the given parameter into a certain direction in cases that can be specified in advance (e.g., they determine the threshold values for triggering rule activation). This intuition relies on the implicit assumption that changes that were appropriate in *another* run on *another* problem are applicable to *this* run on *this* problem.
- **Relative evidence**
  In the case of using relative evidence, parameter values within the *same* run are compared according to the positive/negative effects they produce, and the better values get rewarded. The direction and/or magnitude of the change of the parameter is not specified deterministically, but relative to the performance of other values, i.e., it is necessary to have more than one value present at any given time. As an example, consider an EA using several crossovers with crossover rates adding up to 1.0 and being reset based on the crossover's performance measured by the quality of offspring they create.

8.3.4 What Is the *Scope* of the Change?

As discussed earlier, any change within any component of an EA may affect a gene (parameter), whole chromosomes (individuals), the entire population, another component (e.g., selection), or even the evaluation function. This is the aspect of the scope or level of adaptation [8, 214, 397]. Note, however, that the scope or level is not an independent dimension, as it usually depends on the component of the EA where the change takes place. For example, a change of the mutation step size may affect a gene, a chromosome, or the entire population, depending on the particular implementation (i.e., scheme used), but a change in the penalty coefficients typically affects the whole population. In this respect the scope feature is a secondary one, usually depending on the given component and its actual implementation.

8.3.5 Summary

In conclusion, the main criteria for classifying methods that change the values of the strategy parameters of an algorithm during its execution are:

1. What component/parameter is changed?
2. How is the change made?
3. What evidence is used to make the change?

Our classification is thus three-dimensional. The *component* dimension consists of six categories: representation, evaluation function, variation operators (mutation and recombination), selection, replacement, and population. The other dimensions have respectively three (deterministic, adaptive, self-adaptive) and two categories (absolute, relative). Their possible combinations are given in Table 8.1. As the table indicates, deterministic parameter control with relative evidence is impossible by definition, and so is self-adaptive parameter control with absolute evidence. Within the adaptive scheme both options are possible and are indeed used in practice.

	Deterministic	Adaptive	Self-adaptive
Absolute	+	+	–
Relative	–	+	+

Table 8.1. Refined taxonomy of parameter setting in EAs: types of parameter control along the type and evidence dimensions. The '–' entries represent meaningless (nonexistent) combinations

8.4 Examples of Varying EA Parameters

Here we discuss some illustrative examples concerning all major EA components. For a more comprehensive overview the reader is referred to the classic survey from 1999 [133] and its recent successor [241].

8.4.1 Representation

We illustrate variable representations with the delta coding algorithm of Mathias and Whitley [461], which effectively modifies the encoding of the function parameters. The motivation behind this algorithm is to maintain a good balance between fast search and sustaining diversity. In our taxonomy it can be categorised as an adaptive adjustment of the representation based on absolute evidence. The GA is used with multiple restarts; the first run is used to find an *interim solution*, and subsequent runs decode the genes as distances (*delta values*) from the last interim solution. This way each restart forms a new hypercube with the interim solution at its origin. The resolution of the delta values can also be altered at the restarts to expand or contract the search space. The restarts are triggered when population diversity (measured by the Hamming distance between the best and worst strings of the current population) is not greater than 1. The sketch of the algorithm showing the main idea is given in Fig. 8.2. Note that the number of bits for δ can be increased if the same solution INTERIM is found.

```
BEGIN
  /* given a starting population and genotype-phenotype encoding */
  WHILE (  HD > 1 ) DO
    RUN_GA with k bits per object variable;
  OD
  REPEAT UNTIL (  global termination is satisfied ) DO
    save best solution as INTERIM;
    reinitialise population with new coding;
    /*  k-1 bits as the distance δ to the object value in  */
    /*  INTERIM and one sign bit */
    WHILE (  HD > 1 ) DO
      RUN_GA with this encoding;
    OD
  OD
END
```

Fig. 8.2. Outline of the delta coding algorithm

8.4.2 Evaluation Function

Evaluation functions are typically not varied in an EA because they are often considered as part of the problem to be solved and not as part of the problem-solving algorithm. In fact, an evaluation function forms the bridge between the two, so both views are at least partially true. In many EAs the evaluation function is derived from the (optimisation) problem at hand with a simple transformation of the objective function. In the class of constraint satisfaction problems, however, there is no objective function in the problem definition, cf. Chap. 13. One possible approach here is based on penalties. Let us assume that we have m constraints c_i ($i \in \{1, \ldots, m\}$) and n variables v_j ($j \in \{1, \ldots, n\}$) with the same domain S. Then the penalties can be defined as follows:

$$f(\bar{s}) = \sum_{i=1}^{m} w_i \times \chi(\bar{s}, c_i),$$

where w_i is the weight associated with violating c_i, and

$$\chi(\bar{s}, c_i) = \begin{cases} 1 \text{ if } \bar{s} \text{ violates } c_i, \\ 0 \text{ otherwise.} \end{cases}$$

Obviously, the setting of these weights has a large impact on the EA performance, and ideally w_i should reflect how hard c_i is to satisfy. The problem is that finding the appropriate weights requires much insight into the given problem instance, and therefore it might not be practicable.

The stepwise adaptation of weights (SAW) mechanism, introduced by Eiben and van der Hauw [149], provides a simple and effective way to set these weights. The basic idea behind the SAW mechanism is that constraints that are not satisfied after a certain number of steps (fitness evaluations) must be difficult, and thus must be given a high weight (penalty). SAW-ing changes the evaluation function adaptively in an EA by periodically checking the best individual in the population and raising the weights of those constraints this individual violates. Then the run continues with the new evaluation function. A nice feature of SAW-ing is that it liberates the user from seeking good weight settings, thereby eliminating a possible source of error. Furthermore, the used weights reflect the difficulty of constraints for *the given algorithm* on the *given problem instance* in *the given stage of the search* [151]. This property is also valuable since, in principle, different weights could be appropriate for different algorithms.

8.4.3 Mutation

A large majority of work on adapting or self-adapting EA parameters concerns variation operators: mutation and recombination (crossover). The 1/5 rule of Rechenberg we discussed earlier constitutes a classic example for adaptive mutation step size control in ES. Furthermore, self-adaptive control of mutation step sizes is traditional in ES [257].

8.4.4 Crossover

The classic example for adapting crossover rates in GAs is Davis's adaptive operator fitness. The method adapts the rates of crossover operators by rewarding those that are successful in creating better offspring. This reward is diminishingly propagated back to operators of a few generations back, who helped set it all up; the reward is an increase in probability at the cost of other operators [98]. The GA using this method applies several crossover operators simultaneously within the same generation, each having its own crossover rate $p_c(op_i)$. Additionally, each operator has its local delta value d_i that represents the strength of the operator measured by the advantage of a child created by using that operator with respect to the best individual in the population. The local deltas are updated after every use of operator i. The adaptation mechanism recalculates the crossover rates periodically redistributing 15% of the probabilities biased by the accumulated operator strengths, that is, the local deltas. To this end, these d_i values are normalised so that their sum equals 15, yielding d_i^{norm} for each i. Then the new value for each $p_c(op_i)$ is 85% of its old value and its normalised strength:

$$p_c(op_i) = 0.85 \cdot p_c(op_i) + d_i^{norm}.$$

Clearly, this method is adaptive based on relative evidence.

8.4.5 Selection

Most existing mechanisms for varying the selection pressure are based on the so-called **Boltzmann** selection mechanism, which changes the selection pressure during evolution according to a predefined cooling schedule [279]. The name originates from the Boltzmann trial from condensed matter physics, where a minimal energy level is sought by state transitions. Being in a state i the chance of accepting state j is

$$P[\text{accept } j] = \begin{cases} 1 & \text{if } \; E_i \geq E_j, \\ \exp\left(\frac{E_i - E_j}{K_b \cdot T}\right) & \text{if } \; E_i < E_j, \end{cases}$$

where E_i, E_j are the energy levels, K_b is a parameter called the Boltzmann constant, and T is the temperature. This acceptance rule is called the Metropolis criterion.

We illustrate variable selection pressure in the survivor selection (replacement) step by **simulated annealing** (SA). SA is a generate-and-test search technique based on a physical, rather than a biological, analogy [2, 250]. Formally, however, SA can be envisioned as an evolutionary process with population size of 1, undefined (problem-dependent) representation and mutation, and a specific survivor selection mechanism. The selective pressure changes during the course of the algorithm in the Boltzmann style. The main cycle in SA is given in Fig. 8.3.

In this mechanism the parameter c_k, the temperature, decreases according to a predefined scheme as a function of time, making the probability of accepting inferior solutions smaller and smaller (for minimisation problems). From an evolutionary point of view, we have here a (1+1) EA with increasing selection pressure.

8.4.6 Population

An innovative way to control the population size is offered by Arabas et al. [11, 295] in their GA with variable population size (GAVaPS). In fact, the population size parameter is removed entirely from GAVaPS, rather than

```
BEGIN
  /* given a current solution i ∈ S */
  /* given a function to generate the set of neighbours N_i of i */
  generate j ∈ N_i;
  IF (f(i) < f(j)) THEN
    set i = j;
    ELSE
      IF ( exp((f(i) - f(j)) / c_k) > random[0,1)) THEN
        set i = j;
      FI
    ESLE
  FI
END
```

Fig. 8.3. Outline of the simulated annealing algorithm

adjusted on-the-fly. Certainly, in an evolutionary algorithm the population always has a size, but in GAVaPS this size is a derived measure, not a controllable parameter. The main idea is to assign a lifetime to each individual when it is created, and then to reduce its remaining lifetime by one in each consecutive generation. When the remaining lifetime becomes zero, the individual is removed from the population. Two things must be noted here. First, the lifetime allocated to a newborn individual is biased by its fitness: fitter individuals are allowed to live longer. Second, the expected number of offspring of an individual is proportional to the number of generations it survives. Consequently, the resulting system favours the propagation of good genes.

Fitting this algorithm into our general classification scheme is not straightforward because it has no explicit mechanism that sets the value of the population size parameter. However, the procedure that implicitly determines how many individuals are alive works in an adaptive fashion using information about the status of the search. In particular, the fitness of a newborn individual is related to the fitness of the present generation, and its lifetime is allocated accordingly. This amounts to using relative evidence.

8.4.7 Varying Several Parameters Simultaneously

Mutation, crossover, and population size are all controlled on-the-fly in the GA "without parameters" of Bäck et al. in [25]. Here, the self-adaptive mutation from [17] (Sect. 8.4.3) is adopted without changes, a new self-adaptive technique is invented for regulating the crossover rates of the individuals, and the GAVaPS lifetime idea (Sect. 8.4.6) is adjusted for a steady-state GA model. The crossover rates are included in the chromosomes, much like the mutation rates. If a pair of individuals is selected for reproduction, then their individual crossover rates are compared with a random number $r \in [0, 1]$, and an individual is seen as ready to mate if its $p_c > r$. Then there are three possibilities:

1. If both individuals are ready to mate then uniform crossover is applied, and the resulting offspring is mutated.
2. If neither is ready to mate then both create a child by mutation only.
3. If exactly one of them is ready to mate, then the one not ready creates a child by mutation only (which is inserted into the population immediately through the steady-state replacement), the other is put on hold, and the next parent selection round picks only one other parent.

This study differs from those discussed before in that it explicitly compares GA variants using only one of the (self-)adaptive mechanisms and the GA applying them all. The experiments show remarkable outcomes: the completely (self-)adaptive GA wins, closely followed by the one using only the adaptive population size control, and the GAs with self-adaptive mutation and crossover are significantly worse.

8.5 Discussion

Summarising this chapter, a number of things can be noted. First, parameter control in an EA can have two purposes. It can be done to find good parameter values for the EA at hand. Thus, it offers the same benefits as parameter tuning, but in an on-line fashion. From this perspective tuning and control are two different approaches to solving the same problem. Whether or not one is preferable over the other is an open question with very little empirical evidence to support an answer. Systematic investigations are particularly difficult here because of methodological problems. The essence of these problems is that a fair comparison of the extra computational costs (learning overhead) and the performance gains is hard to define in general.

The other motivation for controlling parameters on-the-fly is the assumption that the given parameter can have a different 'optimal' value in different stages of the search. If this holds, then there is simply no optimal static parameter value; for good EA performance one must vary this parameter. From this perspective tuning and control are not the same, control offers a benefit that tuning cannot.

The second thing to remark is that making a parameter (self-)adaptive does not necessarily mean that we obtain an EA with fewer parameters. For instance, in GAVaPS the population size parameter is eliminated at the cost of introducing two new ones: the minimum and maximum lifetime of newborn individuals. If the EA performance is sensitive to these new parameters then such a parameter replacement can make things worse. This problem also occurs on another level. One could say that the procedure that allocates lifetimes in GAVaPS, the probability redistribution mechanism for adaptive crossover rates (Sect. 8.4.4), or the function specifying how the σ values are mutated in ES (Eq. (4.4)) are also (meta)parameters. It is in fact an assumption that these are intelligently designed and their effect is positive. In many cases there are more possibilities, that is, possibly well-working procedures one can design. Comparing these possibilities implies experimental (or theoretical) studies very much like comparing different parameter values in a classical setting. Here again, it can be the case that algorithm performance is not so sensitive to details of this (meta)parameter, which fully justifies this approach.

Third, it is important to note that the examples in the foregoing sections, while serving as good illustrations of various aspects of parameter control, do not represent the state of the art in 2014. There has been much research into parameter control during the last decade. It has been successfully applied in various domains of metaheuristics, including Evolution Strategies [257], Genetic Algorithms [162, 291], Differential Evolution [349, 280], and Particle Swarm Optimization [473]. Furthermore, there are several noteworthy contributions to the techniques behind parameter control. These range from inventive ideas that need further elaboration, like applying self-organised criticality [266], self-adaptation of population level parameters (e.g., population size) [144] or tuning the controllers to a problem instance [242], to generally applicable mechanisms including adaptive pursuit strategies for operator allocation [429], the Compass method [286] or ACROMUSE [291].

These and other valuable contributions to the field provide more and more evidence about the possible benefits and accumulate the knowhow of successful parameter control. Although the field is still in development, we can identify some trends and challenges. The research community seems to converge on the idea that successful parameter control must take into account two types of information regarding the evolutionary search: data about fitness and population diversity. However, there is a wide variety of approaches to how exactly we should define these types of information; for instance, there are many different ways to define diversity. A very promising approach was put forward recently by McGinley et al. based on the idea of considering the diversity of the fittest part of the population (the 'healthy' individuals) instead of the whole population's diversity [291]. Another agreement on a conceptual level is that a control mechanism is only successful if it appropriately balances exploration and exploitation. But here again, there is no generally accepted definition of these notions, indicating the need for more research [142, 91]. Perhaps one of the biggest obstacles that hinders widespread adoption of existing parameter control techniques is the 'patchwork problem'. The problem here is the lack of generally applicable methods for controlling EA parameters. There are numerous techniques to control mutation, quite a lot for controlling recombination, several ways to adjust population size and a handful for changing selection pressure on-the fly. To build an EA with all parameters controlled, one needs to pick some method for each parameter thus creating a potpourri or patchwork with no solid evidence indicating how it all will work together.

Finally, let us place the issue in a larger perspective of parameter setting in EAs [273]. Over recent decades the EC community shifted from believing that EA performance is to a large extent independent from the given problem instance to realising that it is. In other words, it is now acknowledged that EAs need more or less fine-tuning to specific problems and problem instances. Ideally, this should be automated and advanced (search) algorithms should determine the best EA setting, instead of conventions and the users' intuitions. For the case of doing this in advance, before the EA run there are several powerful algorithms developed over the last ten years, see Section 7.6 and [145]. To put it optimistically, the tuning problem is now solved, and the community of EA researchers and practitioners can adopt tuning as part of the regular workflow. However, the picture is completely different for parameter control. As outlined above, this field is still in its infancy, requiring fundamental research into the most essential concepts (diversity, exploration, etc.) as well as algorithmic development towards good control strategies and some unification (solution of the patchwork problem). To this end, we can recommend the recent overview of Karafotias et al. [241] that identifies current research trends and provides suggestions for important research directions.

For exercises and recommended reading for this chapter, please visit `www.evolutionarycomputation.org`.

9

Working with Evolutionary Algorithms

In this chapter we discuss the practical aspects of using EAs. Working with EAs often means comparing different versions experimentally, and we provide guidelines for doing this, including the issues of algorithm performance measures, statistics, and benchmark test suites. The example application (Sect. 9.4) is also adjusted to the special topics here; it illustrates the application of different experimental practices, rather than EA design.

9.1 What Do You Want an EA to Do?

Throughout this book so far, we have seemingly never considered this issue: "What do you want an EA to do?". The reason is that we tacitly assumed the trivial answer: "I want the EA solve my problem". Many of the subjects treated in this chapter concern specific interpretations and refinements of this answer, and it will become clear that different objectives imply different ways of designing and working with an EA.

A good first step is to examine the given problem context. We can roughly distinguish two main types of problems:

- design (one-off) problems
- repetitive problems, including on-line control problems as special cases

As an example of a design problem, let us consider the optimisation of extensions to an existing road network to meet new demands. This is most certainly a highly complex multiobjective optimisation problem, subject to many constraints. Computer support for this problem requires an algorithm that creates *one* excellent solution at least *once*. In this context the quality of the solution is of utmost importance, and other aspects of algorithm performance are secondary. For instance, since the time scale of the whole project spans years, the algorithm does not have to be fast. It can be given months of computing time, perhaps performing several runs and keeping the best result, if this helps in achieving superior solution quality. The algorithm does not need to be generally applicable either. The present problem most probably contains very specific aspects, hindering reapplication of the algorithm to other problems. Furthermore, a similar problem will occur as part of a similar project allowing enough time to develop a good EA for that problem.

Repetitive problems form a counterpoint to design problems. As an illustration, consider a domestic transportation firm, having dozens of trucks and drivers that need to be given a daily schedule every morning. The schedule should contain a pick-up and delivery plan, plus the corresponding route description for each truck and driver. For each of them, this is just a TSP problem (probably with time windows), but the optimisation criteria and the constraints must be taken across the whole firm, together making the actual problem very complex. Depending on the type of business, the data and requirements for a day's schedule might become available weeks, days, but maybe only hours before the schedules are to be handed out to the drivers. In any case, the dispatcher must provide a schedule every morning to every available driver. Suitable computer support for this problem, an EA in our case, must be able to find *good* solutions *quickly* and be able to do this *repeatedly* for *different instances* of the problem (i.e., with different data and requirements every day). The implications for the algorithm are radically different from those in the case of a design problem. The balance in the speed versus quality trade-off is clearly towards speed. Solutions must be good, e.g., better than hand-made ones, but not necessarily optimal. Speed, however, is crucial. For example, it could be required that the time between feeding the data into the system and receiving the schedules does not exceed 30 minutes. Closely related to this issue, it is important that the algorithm performance is stable. Since an EA is a stochastic algorithm, the quality of end solutions over a number of runs shows a certain variance. For a design problem we typically have the time to perform many runs and select the best solution. Therefore it is not a problem if some runs terminate with bad results, as long as others end with good solutions. For repetitive problems, however, we might only have time for one run. To reduce the probability of really bad runs, we need a consistent EA to keep the variance of end solution quality as low as possible. Finally, for repetitive problems the widescale applicability of the algorithm is also important as the system will be used under various circumstances. In other words, the algorithm will be run on different problem instances.

On-line control problems can also be seen as repetitive problems with extremely tight time constraints. To remain in the transportation context, we can think of traffic light optimisation. In particular, let us consider the task of optimising a controller to set the green times of a single crossing with four crossroads. We assume that each of the crossroads has sensors embedded in the road surface that continuously monitor traffic approaching the crossing.[1] This sensory information is sent to the traffic light controller, a piece of software running on a special device at the crossing. The task of this controller

[1] This is common in many countries, and standard in the Netherlands.

is to calculate the green times for each of the roads in such a way that the throughput of vehicles is maximised. It is important to note that an EA can be used for this problem in two completely different ways. First, we can use an EA *to develop a controller*, based on simulations, which is then deployed at the crossing in question. This type of application was mentioned in Sect. 6.4 on genetic programming. The other way is to have an EA that *is the controller*, and this is what we have in mind here. The most important requirement here is, of course, speed. A controller is working in on-line mode, and it has to cope with streaming sensory information and needs to control the traffic lights in real time. The speed requirements are given in wall-clock time: the length of one full cycle[2] is typically a few minutes, and this time must be enough to calculate the green times for the following cycle. This can be very demanding for an EA that works with a whole population of candidate solutions and needs quite a few generations to evolve a good result. Fortunately, by the nature of traffic flows, the situation does not change very rapidly, which also holds for many other control problems. This means that the consecutive problem instances are rather similar to each other, and therefore it can be expected that the corresponding near-optimal solutions are similar as well. This motivates an EA that keeps the best solutions from previous runs and uses them in the new run. The second requirement, similar to repetitive problems, is a small variance in end solution quality. The third one is that the controller (the EA) must be very fault-tolerant and robust. This means that noise in the data (measurement errors of the sensors) or missing data (breakdown of a sensor) must not have a critical effect. The system, and the EA, must keep working and delivering the best possible results under the given circumstances.

Finally, let us mention a different but important context for working with EAs: academic research. The considerations above apply in an application-oriented situation, and it can be argued that making good applications is one of the major goals of the whole evolutionary computing field. However, an examination of the EC literature soon reveals that a huge majority of papers in scientific journals, conference proceedings, or monographs are ignorant of such concrete application-related issues. Scientific research apparently has its own dynamics, goals, methodologies, and conventions. Some of these arise from the fact that EAs can exhibit complex behaviours and emergent phenomena that are interesting per se, and developing a solid theoretical understanding may yield insight into real biological evolution. This chapter would not be complete without paying attention to working with EAs in an academic environment.

The objective in many experimental research papers, implicitly or explicitly, is to show that some EA is better than other EAs or their competitors – at least for some 'interesting' problems. This objective is typically not placed into an application context. The requirements of the algorithm are therefore

[2] One cycle is defined as the time between two consecutive turn-to-green moments of traffic light no. 1.

not inferred from what we want it to do; rather, they are based on conventions or ad hoc choices. Typical goals for academic experimentation are:

- Get a good solution for a given problem, e.g., challenging combinatorial optimisation.
- Show that EC is applicable in a (possibly new) problem domain.
- Show that an EA with some newly invented feature is better than some benchmark EA.
- Show that EAs outperform traditional methods on some relevant problems.
- Find best setup for parameters of a given EA, in particular, get data on the impact of varying some EA component, e.g., the population size.
- Obtain insights into algorithm behaviour, e.g., the interaction between selection and variation.
- See how an EA scales-up with problem size.
- See how the performance is influenced by parameters of the problem *and* the algorithm.

While these goals are different among themselves, and academic experimental research is apparently different from application-oriented work, there are general issues for all of these cases. The most prominent issue present in all experimental work is the objective of assessing algorithm performance.

9.2 Performance Measures

Assessing the quality of an evolutionary algorithm commonly implies experimental comparisons between the given EA and other evolutionary or traditional algorithms. Even if showing the superiority of some EA is not the main goal, parameter tuning for good performance still requires experimental work to compare different algorithm variants.

Such comparisons always assume the use of some algorithm performance measures, since claims about ranking algorithms are always meant in terms of their relative performances rather than, for instance, code length or readability. Because EAs are stochastic, performance measures are statistical in nature, meaning that a number of experiments need to be conducted to gain sufficient experimental data, as noted in Sect. 7.5. In the following we discuss three basic performance measures:

- success rate
- effectiveness (solution quality)
- efficiency (speed)

Additionally, we discuss the use of progress curves, i.e., plots of algorithm behaviour against time.

9.2.1 Different Performance Measures

In quite a few cases, experimental research concerns problems where either the optimal solution can be recognised, which is typical in academia, or a criterion for sufficient solution quality can be given, as in many practical applications. In these cases one can easily define a success criterion: finding a solution of the required quality, and the **success rate** (SR) measure can be defined as the percentage of runs where this happens. For problems where the optimal solutions cannot be recognised, the SR measure cannot be used in theory. This is the case if the optimum of the objective function is unknown, or if perhaps not even a lower/upper bound is available. Nevertheless, a success criterion in the *practical* sense can often be given even in these cases. For example, think of a university timetabling problem. The theoretical optimum for any given year is surely unknown here. However, one could use last year's timetable, or the one made by hand as benchmark and declare that a run ending with a timetable beating the benchmark by 10% is a success. Practical success criteria can also be used even in cases when the theoretical optimum is known, but the user does not require this optimum. For instance, it might be sufficient if we have a solution with an error less than a given $\epsilon > 0$.

The **mean best fitness** measure (MBF) can be defined for any problem that is tackled with an EA – at least for any EA using an explicit fitness measure (thus excluding, for instance, interactive evolution applications, Sect. 14.1). For each run of a given EA, we record the fitness of the best individual at termination. The MBF is the average of these values over all runs.

Note that although SR and MBF are related, they are different, and there is no general advice on which one to use for algorithm comparison. The difference between the two measures is rather obvious: SR cannot be defined for some problems, while the MBF is always a valid measure. Furthermore, all possible combinations of low or high SR and MBF values can occur. For example, low SR and high MBF is possible and indicates a good approximizer algorithm: it gets close consistently, but seldom really makes it. Such an outcome could motivate increasing the length of the runs, hoping that this allows the algorithm to finish the search. An opposite combination of a high SR and low MBF is also possible, indicating a 'Murphy algorithm': if it goes wrong, it goes very wrong. That is, those few runs that terminate without an (optimal) solution end in a disaster, with a very bad best fitness value deteriorating MBF. Clearly, whether the first or the second type of algorithm behaviour is preferable depends on the problem. As mentioned above, for a timetabling problem the SR measure might not be meaningful, so one should be interested in a high MBF. To demonstrate the other situation, think of solving the 3-SAT problem with the number of unsatisfied clauses as fitness measure. In this case a high SR is pursued, since the MBF measure – although formally correct – is useless because the number of unsatisfied clauses at termination says, in general, very little about how close the EA got to a solution. Notice, however, that the particular application objectives (coming from the original problem-solving context) might necessitate a refinement of this picture. For instance, if the 3-SAT problem to be solved represents a practical problem, with some tolerance for a solution, then measuring MBF and striving for a good MBF value might be appropriate.

In addition to the mean best fitness calculated over a number of runs, in specific cases one might be interested in the best-ever or the worst-ever fitness. As discussed above, for design problems, the best-ever fitness is more appropriate than MBF, since *one* excellent solution is all that is required. For repetitive problems the worst-ever fitness can be interesting, as it can be used for studying worst-case scenarios and can help to establish statistical guarantees on solution quality.

It is important to note that for both SR and MBF, it is assumed that they are measured using an a priori specified limit of computational efforts. That is, SR and MBF always reflect performance within a fixed maximum amount of computing. If this maximum is changed, the ranking of algorithms might change as well. This is illustrated in Fig. 9.1, which shows a 'tortoise and hare' situation, where algorithm A (the hare) shows rapid progress, and in the case of limited time it beats algorithm B (the tortoise). In turn algorithm B outperforms algorithm A if given more time. Summarising, SR and MBF are performance measures for an algorithm's **effectiveness**, indicating how far can it get within a given computational limit.

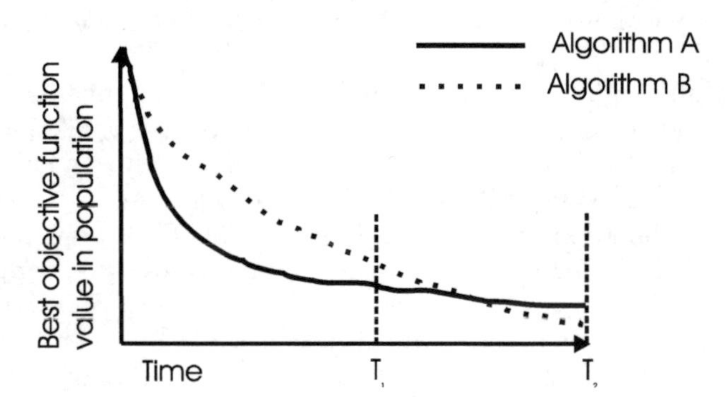

Fig. 9.1. Comparing algorithms A and B by after terminating at time T_1 and T_2 (for a minimisation problem). Algorithm A clearly wins in the first case, while B is better in the second one

The complementary approach is to specify when a candidate solution is satisfactory and measure the amount of computing needed to achieve this solution quality. Roughly speaking, this is the issue of algorithm **efficiency** or speed. Speed is often measured in elapsed computer time, CPU time, or user time. However, these measures depend on the specific hardware, operating system, compiler, network load, and so on, and therefore are ill-suited for reproducible research. In other words, repeating the same experiments, possibly elsewhere, may lead to different results. For generate-and-test-style algorithms, such as EAs, a common way around this problem is to count the number of points visited in the search space. Since EAs immediately evaluate each newly generated candidate solution, this measure is usually expressed as the number of fitness evaluations. Of necessity, because of the stochastic nature of EAs, this is always measured over a number of independent runs, and the **average number of evaluations to a solution** (AES) is used. It is important to note that the average is only taken over the successful runs (that is, 'to a solution'). Sometimes the average number of evaluations to termination measure is used instead of the AES, but this has clear disadvantages. Namely, for runs finding no solutions, the specified maximum number of evaluations will be used when calculating this average. This means that the values obtained will depend on how long the unsuccessful runs are allowed to continue. That is, this measure mixes the AES and the SR measures, and the outcome figures are hard to interpret.

Using the AES measure generally gives a fair comparison of algorithm speed, but its usage can be disputed, or even misleading in some cases:

1. First, if an EA uses 'hidden labour', for instance, some local search heuristics incorporated in the mutation operator. The extra computational efforts may increase performance, but are invisible to the AES measure.
2. Second, if some evaluations take longer than others. For instance, if a repair mechanism is applied, then evaluations invoking this repair take much longer. One EA with good variation operators might proceed by chromosomes that do not have to be repaired, while another EA may need a lot of repair. The AES values of the two may be close, but the second EA would be much slower, and this is not an artifact of the implementation.
3. Third, if evaluations can be done very quickly compared with executing other steps in the EA cycle.[3] Then the AES does not truly reflect algorithm speed as other components of the EA have a relatively large impact.

An additional problem with AES is that it can be difficult to apply for comparing an EA with search algorithms that do not work in the same search space, in the same fashion. An EA iteratively improves complete candidate solutions, so each elementary search step consists of the creation and testing of one new candidate solution. However, a constructive search algorithm works in the space of partial solutions (including the complete ones through which an EA is searching), so one elementary search step consists of extending the current solution. In general, counting the number of elementary search steps is misleading unless the nature of those steps is the same. A possible treatment for this, and also for the hidden labour problem, is to compare the scale-up behaviour of the algorithms. This requires a problem that is scalable, i.e., its

[3] Typically this is not the case, and around 70–90% of the time is spent on fitness evaluations.

size can be changed. The number of variables is a natural scaling parameter for many problems. Two different types of methods can then be compared by plotting their own speed measure figures against the problem size. Even though the measures used in each curve are different, the steepness information is a fair basis for comparison: the curve that grows at a higher rate indicates an inferior algorithm (Fig. 9.2). A great advantage of this comparison is that it can also be applied to plain running times (e.g., CPU times), not only to the number of abstract search steps. As discussed above, there are important arguments against using running times themselves for comparisons. However, the scale-up curves of running times do give a fair comparison without those drawbacks.

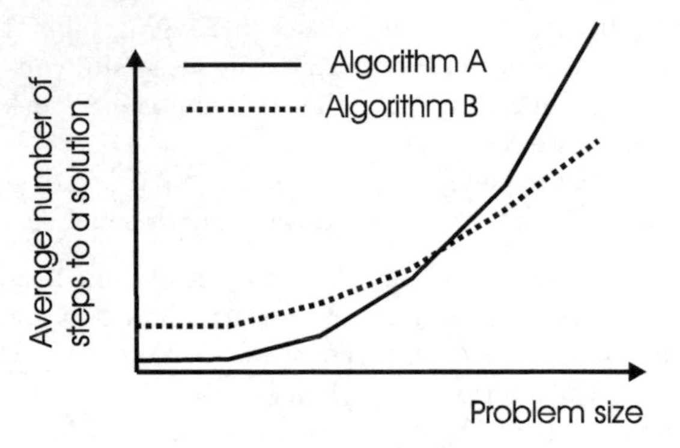

Fig. 9.2. Comparing algorithms A and B by their scale-up behaviour. Algorithm B can be considered preferable because its scale-up curve is less steep

Success percentages and run lengths can be meaningfully combined into a measure expressing the amount of processing required to solve a problem with a given probability [252, Chap. 8]. This measure (defined for generational EAs and used frequently in GP) depends on the population size and the number of generations as tuneable quantities. The probability $Y(\mu, i)$ that a given run with population size μ hits a solution for the first time in generation i is estimated by observed statistics, which require a substantial number of runs. Cumulating these estimations, we can calculate the probability $P(\mu, i)$ that a given generation i will contain a solution (found in a generation $j \leq i$), and hence the probability that generation i finds a solution at least once in R runs as $1 - (1 - P(\mu, i))^R$. Then the number of independent runs needed to find a solution by generation i with a probability of z is

$$R(\mu, i, z) = \left\lceil \frac{log(1 - z)}{log(1 - P(\mu, i))} \right\rceil, \tag{9.1}$$

where $\lceil \ \rceil$ is the ceiling function. Being a function of the population size, this measure can give information on how to set μ. For instance, after collecting enough data with different settings, the total amount of processing, that is, the number of fitness evaluations, needed to find a solution with a probability of z by generation i using a population size μ is $I(\mu, i, z) = \mu \cdot i \cdot R(\mu, i, z)$. Notice that the dependence on μ is not the crucial issue here; in fact, any algorithm parameter p can be used in an analogous way to estimate for $R(p, i, z)$.

Another alternative to AES, especially in cases where one cannot specify satisfactory solution quality in advance, is the pace of progress to indicate algorithm speed. Here the best (or alternatively the worst or average) fitness value of the consecutive populations is plotted against a time axis – typically the number of generations or fitness evaluations (Fig. 9.1). Clearly, such a plot provides much more information than the AES, and therefore it can also be used when a clear success criterion is available. In particular, a progress plot can help rank two algorithms that score the same on AES. For example, progress curves might disclose that algorithm A has achieved the desired quality halfway through the run. Then the maximum number of evaluations might be decreased and the competition redone. The chance is high that algorithm A keeps its performance, e.g., its MBF, at lower costs and algorithm B does not, thus a well-motivated preference can be formulated. Another possible difference between progress curves of algorithms can be the steepness towards the end of the run. If, for instance, curve A has already flattened out, but curve B did not, one might extend the runs. The chance is high that B will make further progress in the extra time, but A will not; thus again, the two algorithms can be distinguished.

A problem with using such progress plots is that it is hard to use them in a statistical way. Averaging the data of, say, 100 runs and only drawing the average plot can hide interesting effects by smoothening them out. Overlaying all curves forms an alternative, but has obvious disadvantages: it might result in a chaotic figure with too much black ink and no visible structure. A practical solution is depicting a typical curve, that is, one single plot that is representative for all others. This option might not have a solid statistical basis, but it can deliver the most information when used with care.

9.2.2 Peak Versus Average Performance

For some, but not all, performance measures, there is an additional question of whether one is interested in peak performance, or average performance, considered over all these experiments. In evolutionary computing it is typical to suggest that algorithm A is better than algorithm B if its average performance is better. In many applications, however, one is often interested in the best solution found in X runs or within Y hours/days/weeks (peak performance), and the average performance is not that relevant. This is typical, for instance, in design problems as discussed in Section 9.1. In general, if there is time for more runs on the given problem and the final solution can be selected from the best solutions of these runs, then peak performance is more relevant than average performance.

We have a different situation if a problem-solving session only allows time for one run that must deliver the solution. This might be the case if a computationally expensive simulation is needed to calculate fitness values or for a real-time application, like repetitive and on-line control problems. Here, an algorithm with high average performance and small standard deviation is the best option, since it carries the lowest risk of missing the only chance we have.

It is interesting to note that academic experimental EC research falls in the first category – there is always time to perform more runs on any given set of test problems. In this light, it is strange that the huge majority of experimental EC research is comparing average performances of algorithms. This might be because researchers do not consider the differences between design and repetitive problems, and do not realise the different implications for the requirements of the problem-solving algorithm. Instead, it seems, they simply assume that the EA will be used in the repetitive mode.

Next we consider an example to show how the interpretation of figures concerning averages and standard deviations can depend on application objectives. In EC it is common to express preferences for algorithms with better averages for a given performance measure, e.g., higher MBF or lower AES, especially if a better average is coupled to a lower standard deviation. This attitude is never discussed, but it is less self-evident than it might seem. Using the timetabling example, let us assume that two algorithms are compared based on 50 independent runs and the resulting MBF values that are given in Fig. 9.3. Given these results, it could be tempting to conclude that algorithm A is better, because of the slightly higher MBF, and the more consistent behaviour (that is, lower variation in best fitness values at termination). This is indeed a sound argument in the case of a repetitive application, for instance, if a team of hospital employees must be scheduled every morning, based on fresh data and constraints. Notice, however, that six runs of algorithm B terminated with a solution quality that algorithm A never achieved. Therefore in a design application algorithm B is preferable, because of the higher chance of delivering a better timetable. Making a university timetable would fall into this category, since it has to be made only once a year, and the data is available weeks, perhaps months, before the timetable must be effective. This discussion of performance measures is not exhaustive, but it illustrates the point that for a sound comparison it is necessary to specify the objectives of the algorithms in light of some problem-solving context and to derive the performance measures used for comparison from these objectives.

Finally, let us pay some attention to using statistics. It is clear that by the stochastic nature of EAs only statistical statements about behaviour are possible. Typically, averages and standard deviations supply the necessary basis for claims about (relative) performance of algorithms. In a few cases these can be considered as sufficient, but in principle it is possible that two (or more) series of runs deliver data that are statistically indistinguishable, i.e., may come from the same distribution, and the differences are due to random effects. This means that the two series, and the behaviour of the two

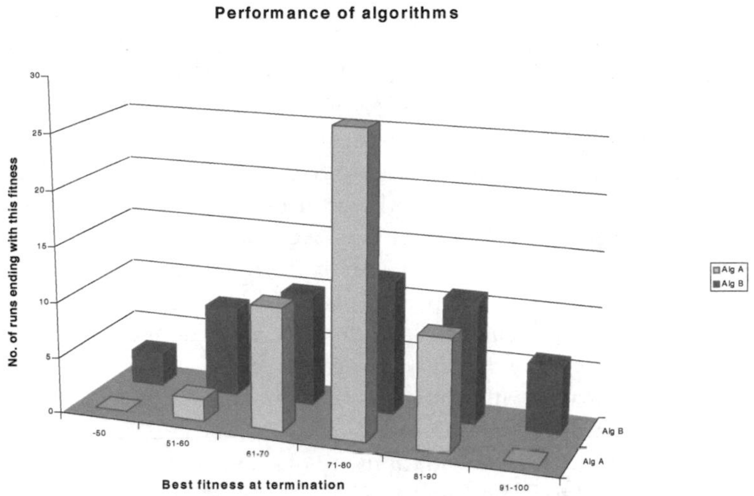

Fig. 9.3. Comparing algorithms by histograms of the best found fitness values

EAs behind these series, should be considered as statistically identical, and claims about one being better are ill-founded. It is important to recall that the mean and standard deviation of any given observable are only *two* values by which we try to describe the whole set of data. Consequently, considering only the standard deviations often cannot eliminate the possibility that any observed difference is only a result of randomness.

Good experimental practice therefore requires the use of specific tests to establish whether observed differences in performance are truly statistically significant. A popular method for this purpose is the two-tailed t-test, which gives an indication about the chance that the values came from the same underlying distribution. The applicability of this test is subject to certain conditions, for instance, that the data are normally distributed, but in practice it proves rather robust and is often used without verifying these conditions. When more than two algorithms are being compared, it is suggested that an analysis of variance (ANOVA) test be performed. This uses all of the data to calculate a probability that any observed differences are due to random effects, and should be performed *before* comparing algorithms pairwise. More sophisticated variants also exist: for instance, if we want to investigate the effects of two parameters (say, population size and mutation rate), then we can perform a two-way ANOVA, which simultaneously analyses the effects of each parameter and their interactions.

If we have limited amounts of data, or our results are not normally distributed, then using t-tests and ANOVA is not appropriate. For example, if we are comparing MBF values for a number of algorithms, with $SR < 1$ for some (but not all) of them, then our data will almost certainly not be normally distributed, since there is a fixed upper limit to the MBF value defined by the problem's optimum. In this type of cases, it is better to use the equivalent rank-based non-parametric test, noting that it is less likely to show a difference, since it makes fewer assumptions about the nature of the data.

Unfortunately, the present experimental EC practice seems rather unaware of the importance of statistics. This is a great shame, since there are many readily available software packages for performing these tests, so there is no excuse for not performing a proper statistical analysis of results. However this problem is easily remediable. There are any number of excellent books on statistics that deal with these issues, aimed at experimental sciences, or business and management courses, see for instance [290, 472]. The areas of concern are broadly known as hypothesis testing, and experimental design. Additionally, a wealth of information and on-line course material can be found by entering these terms into any Internet search engine.

9.3 Test Problems for Experimental Comparisons

In addition to the issue of performance measures, experimental comparisons between algorithms require a choice of benchmark problems and problem instances. We distinguish three different approaches:

1. Using problem instances from an academic benchmark repository.
2. Using problem instances created by a problem instance generator.
3. Using real-life problem instances.

9.3.1 Using Predefined Problem Instances

The first option amounts to obtaining prepared problem instances that are freely available from Web-based repositories, monographs, or other printed literature. In the history of EC some objective functions had a large impact on experimental studies. For instance, the so-called De Jong test suite, consisting of five functions has long been very popular [102]. This test suite was carefully designed to span an interesting variety of fitness landscapes. However, both computing power and our understanding of EAs have advanced considerably since the 1970s. Consequently, a modern study that only showed results on these functions and then proceeded to make general claims would not be considered methodologically sound. Over the last decade other functions have been added to the 'obligatory' list and are used frequently, such as the Ackley, Griewank, and Rastrigin functions, just to name the most popular ones. Obviously, new functions pose new challenges to evolutionary algorithms, but the improvement is still rather quantitive. To put it plainly: How much better is a claim based on ten test landscapes than one only using the five De Jong functions? There is, of course, a straightforward solution to this problem by limiting the scope of statements about EA performance and restricting it to *the functions used* in the comparative experiments. Formally, this is a sound option, but in practice these careful refinements can easily skip the reader's attention. Additionally, the whole EC community using the same test suite can lead to overfitting new algorithms to these test functions. In other words, the community will not develop better and better EAs over the years, but only better and better EAs for these problems!

Another problem with the present practice of using particular objective functions or fitness landscapes is that these functions do not form a systematically searchable collection. That is, using 15 such functions will deliver 15 data points without structure. Unfortunately, although we have some ideas about the sorts of features that make problems hard for EAs, we do not currently possess the tools to divide these functions into meaningful categories, so it is not possible to draw conclusions on the relationship between characteristics of the problem (the objective function) and the EA behaviour. A deliberate attempt by Eiben and Bäck [130] in this direction failed in the sense that the EA behaviour turned out to be inconsistent with the borders of the test function categories. In other words, the EAs showed different behaviours within one category and similar behaviours on functions belonging to different categories. This example shows that developing a meaningful classification of objective functions or test landscapes is nontrivial because the present vocabulary to describe and to distinguish test functions seems inappropriate to define good categories (see [239] for a good survey of these issues). For the time being this remains a research challenge [135].

Building on cumulative experience in the EC community, for instance that of Whitley et al. [454], Bäck and Michalewicz gave some general guidelines for composing test suites for EC research in [29]. Below we reproduce the main points from their recommendations. The test suite should contain:

1. A few unimodal functions for comparisons of convergence velocity (efficiency), e.g., AES.
2. Several multimodal functions with a *large* number of local optima (e.g., a number growing exponentially with n, the search space dimension). These functions are intended to be representatives of the characteristics that are typical for real-world problems, where the best out of a number of local optima is sought.
3. A test function with randomly perturbed objective function values models a typical characteristic of numerous real-world applications and helps to investigate the robustness of the algorithm with respect to noise.
4. Constrained problems, since real-world problems are typically constrained, and constraint handling is a topic of active research.

5. High-dimensional objective functions, because these are representative of real-world applications. Furthermore, low-dimensional functions (e.g., with $n = 2$) are not suitable representatives of application problems where an evolutionary algorithm would be applied, because they can be solved optimally with traditional methods. Most useful are test functions that are *scalable* with respect to n, i.e., which can be used for arbitrary dimensions.

9.3.2 Using Problem Instance Generators

An alternative to such test landscapes is formed by problem instances of a certain (larger) class, for instance, operations research problems, constrained problems or machine-learning problems. The related research communities have developed their collections, like the OR library `http://www.ms.ic.ac.uk/info.html`, the constraints archive at `http://www.cs.unh.edu/ccc/archive`, or the UCI Machine Learning Repository at `http://www.ics.uci.edu/~mlearn/MLRepository.html`. The advantage of such collections is that the problem instances are interesting in the sense that many other researchers have investigated and evaluated them already. Besides, an archive often contains performance reports of other techniques, facilitating direct feedback on one's own achievements.

Over the last few years there has been a growing research interest in using problem instance generators. Using such a generator, which could of course come from an archive, means that problem instances are produced on-the-spot. Generators usually have some problem-specific parameters, for example, the number of clauses and the number of variables for 3-SAT, or the number of variables and the extent of their interaction for NK landscapes [244], and can generate random instances for each parameter value. The advantage of this approach is that the characteristics of the problem instances can be tuned by the generator's parameters. In particular, for many combinatorial problems there is a lot of information available about the location of really hard problem instances, the so-called phase transition, related to the given parameters of the problem [80, 183, 344]. A generator makes it possible to perform a systematic investigation in and around the hardest parameter range. Thus one can create results relating problem characteristics to algorithm performance. An illustration is given in Fig. 9.4. The question "which of the two algorithms is better" can now be refined to "which algorithm is better on which problem instances". On mid-range parameter values (apparently the hardest instances) algorithm B outperforms algorithm A. On the easier instances belonging to low and high parameter values, this behaviour is reversed.

9.3.3 Using Real-World Problems

Testing on real data has the advantages that results can be considered as very relevant viewed from the application domain (data supplier). However, it also has some disadvantages. Namely, practical problems can be overcomplicated.

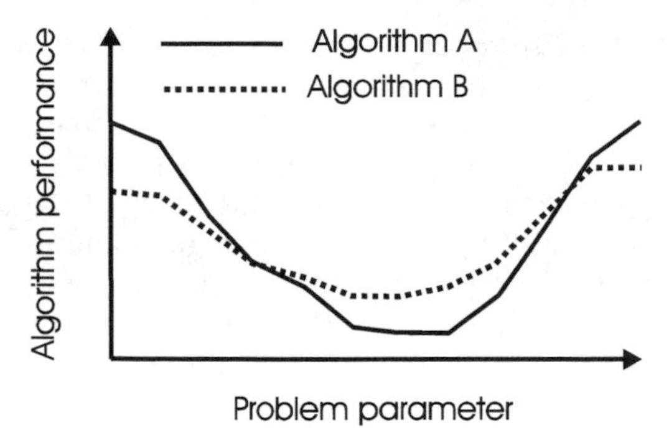

Fig. 9.4. Comparing algorithms on problem instances with a scalable parameter

Furthermore, there can be few available sets of real data, and these data may be commercially sensitive and therefore difficult to publish and to allow others to compare. Last, but not least, there might be so many application-specific aspects involved that the results are hard to generalise. Despite these drawbacks it remains highly relevant to tackle real-world problems as the proof of the pudding is in the eating!

9.4 Example Applications

As mentioned in the introduction to this chapter, instead of presenting two case studies with implementation details, we next describe examples of good and bad practice, in order to illustrate some of our points.

9.4.1 Bad Practice

This section shows a hypothetical example of an experimental study following the template that can be found in many EC publications.[4] In this imaginary case a researcher has invented a new EA feature, e.g., "tricky mutation", and assessed the value of this new feature by running a standard GA and "tricky GA" 20 times independently on each of 10 objective functions chosen from the literature. The outcomes of these experiments proved tricky GA better on seven, equal on one, and worse on two objective functions in terms of SR. On this basis it was concluded that the new feature is indeed valuable.

The main question here is what did we, the EC community, learn from this experience? We did learn a new feature (tricky mutation) and obtained some indication that it might be a promising idea to try in a GA. This can of course justify publishing a paper reporting this; however, there are also many things that we did not learn here, including:

[4] The authors admit that some of their own papers also follow this template.

- How relevant are these results, e.g., are the test functions typical of real-world problems, or important only from an academic perspective?
- What would have happened if a different performance metric had been used, or if the runs had been ended sooner, or later?
- What is the scope of claims about the superiority of the tricky GA?
- Is there a property distinguishing the seven good and two bad functions?
- Are these results generalisable? Alternatively, do some features of the tricky GA make it applicable for other specific problems, and if so which?
- How sensitive are these results to changes in the algorithm's parameters?
- Are the performance differences as measured here statistically significant, or can they be just artifacts caused by random effects?

The next example explicitly addresses some of these issues and therefore forms a showcase for a better, albeit still not perfect, practice.

9.4.2 Better Practice

A better example of how to evaluate the behaviour of a new algorithm takes into account questions such as:

- What type of problem am I trying to solve?
- What would be a desirable property of an algorithm for this type of problem, for example: speed of finding good solutions, reliably locating good solutions, or occasional brilliance?
- What methods currently exist for this problem, and why am I trying to make a new one, i.e., when do they not perform well?

After considering these issues, a particular problem type can be chosen, a careful set of experiments can be designed, and the necessary data to collect can be identified. A typical process might proceed along the following lines:

- inventing a new EA (xEA) for solving problem X
- identifying three other EAs and a traditional benchmark heuristic for problem X in the literature
- asking when and why xEA could be better than the other four methods
- obtaining a problem instance generator for problem X with two parameters: n (problem size) and k (some problem-specific indicator)
- selecting five values for k and five values for n
- generating 100 random problem instances for all 25 combinations
- executing all algorithms on each instance 100 times (the benchmark heuristic is also stochastic)
- recording AES, SR, and MBF values and standard deviations (not for SR)
- identifying appropriate tests based on the data and assessing the statistical significance of results
- putting the program code and the instances on the Web

The advantages of this template with respect to the one in the previous example are numerous:

- The results can be arranged in 3D: that is, as a performance landscape over the (n, k) plane with special attention to the effect of n on scale-up.
- The niche for xEA can be identified, e.g., weak with respect to other algorithms for (n, k) combinations of type 1, strong for (n, k) combinations of type 2, comparable otherwise. Thus the 'when' question can be answered.
- Analysing the specific features and the niches of each algorithm can shed light on the 'why' question.
- A lot of knowledge has been collected about problem X and its solvers.
- Generalisable results are achieved, or at least claims with well-identified scope based on solid data.
- Reproduction of the results, and further research elsewhere, is facilitated.

For exercises and recommended reading for this chapter, please visit `www.evolutionarycomputation.org`.

Part III

Advanced Topics

10

Hybridisation with Other Techniques: Memetic Algorithms

In the preceding chapters we described the main varieties of evolutionary algorithms and described various examples of how they might be suitably implemented for different applications. In this chapter we turn our attention to systems in which, rather than existing as stand-alone algorithms, EA-based approaches are either incorporated within larger systems, or alternatively have other methods or data structures incorporated within them. This category of algorithms is very successful in practice and forms a rapidly growing research area with great potential. This area and the algorithms that form its subject of study are named memetic algorithms (MA). In this chapter we explain the rationale behind MAs, outline a number of possibilities for combining EAs with other techniques, and give some guidelines for designing successful hybrid algorithms.

10.1 Motivation for Hybridising EAs

There are a number of factors that motivate the hybridization of evolutionary algorithms with other techniques. In the following we discuss some of the most salient of these. Many complex problems can be decomposed into a number of parts, for some of which exact methods, or very good heuristics, may already be available. In these cases it makes sense to use a combination of the most appropriate methods for different subproblems.

Overall, successful and efficient general problem solvers do not exist. The rapidly growing body of empirical evidence and some theoretical results, like the No Free Lunch theorem (NFL),[1] strongly support this view. From an EC perspective this implies that EAs do not exhibit the performance as suggested in the 1980s, cf. Fig. 3.8 in Sect. 3.5. An alternative view on this issue is

[1] The NFL is treated in detail in Chap. 16, including a discussion about what it really says. For the present we interpret it as stating that all stochastic algorithms have the same performance when averaged over all discrete problems.

given in Fig. 10.1. The figure considers the possibility that we could combine problem-specific heuristics and an EA into a hybrid algorithm. Furthermore, it is assumed that the amount of problem-specific knowledge is variable and can be adjusted. Depending on the amount of problem-specific knowledge in the hybrid algorithm, the global performance curve will gradually change from roughly flat (pure EA) to a narrow peak (problem-specific method).

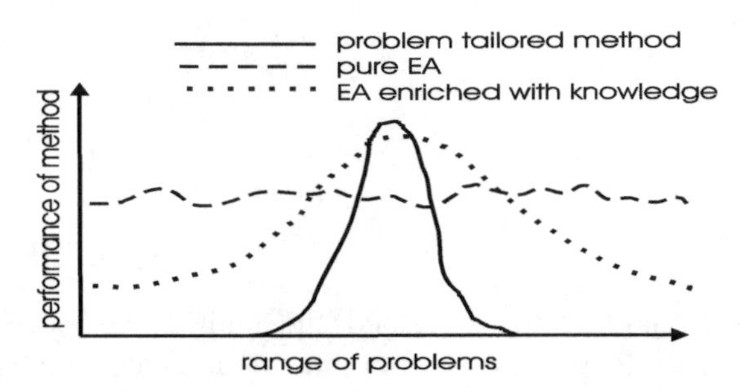

Fig. 10.1. 1990s view of EA performance after Michalewicz [295]

In practice we frequently apply an evolutionary algorithm to a problem where there is a considerable amount of hard-won user experience and knowledge available. In such cases performance benefits can often arise from utilising this information in the form of specialist operators and/or good solutions, provided that care is taken not to bias the search too much away from the generation of novel solutions. In these cases it is commonly experienced that the combination of an evolutionary and a heuristic method – a **hybdrid EA** – performs better than either of its 'parent' algorithms alone. Note, that in this sense, Figure 10.1 is misleading as it does not indicate this effect.

There is a body of opinion that while EAs are very good at rapidly identifying good areas of the search space (exploration), they are less good at the 'endgame' of fine-tuning solutions (exploitation), partly as a result of the stochastic nature of the variation operators. To illustrate this point, as anyone who has implemented a GA to solve the One-Max problem[2] knows, the algorithm is quick to reach near-optimal solutions, but the process of mutation finding the last few bits to change can be slow, since the choice of which genes are mutated is random. A more efficient method might be to incorporate a more systematic search of the vicinity of good solutions by adding a local search improvement step to the evolutionary cycle (in this case, a bit-flipping hill-climber).

A final concept, which is often used as a motivation by researchers in this field, is Dawkins' idea of **memes** [100]. These can be viewed as units of cultural

[2] A binary coded maximisation problem, where the fitness is simply the count of the number of genes set to 1.

transmission, in the same way that genes are the units of biological transmission. They are selected for replication according to their perceived utility or popularity, and then copied and transmitted via interperson communication.

> Examples of memes are tunes, ideas, catch-phrases, clothes fashions, ways of making pots or of building arches. Just as genes propagate themselves in the gene pool by leaping from body to body via sperm or eggs, so memes propagate themselves in the meme pool by leaping from brain to brain via a process which, in the broad sense, can be called imitation [100, pg. 192].

Since the idea of memes was first proposed by Dawkins, it has been extended by other authors (e.g., [57, 70]). From the point of view of the study of adaptive systems and optimisation techniques, it is the idea of memes as agents that can transform a candidate solution that is of direct interest. We can consider the addition of a learning phase to the evolutionary cycle as a form of meme–gene interaction, whereby the problem representation (genotype) is now considered to be 'plastic', and the influence of the learning mechanism (meme) can be thought of as a developmental process.

In the years since 2000 there has been an increasing amount of attention paid to the concept that, rather than acting as fixed learning strategies, memes themselves could be subjected to selection and adaptation according to their perceived usefulness, giving rise to the field that has become known as Adaptive Memetic Algorithms [264, 326, 328]. Section 10.4 describes this progression in more detail.

Extending this perspective beyond local search-evolutionary hybrids, Ong *et al.* consider Memetic Computation as a more general paradigm which uses *"the notion of meme(s) as units of information encoded in computational representations for the purposes of problem solving"*, cf. [327]. In their more general view memes might be represented as *"decision trees, artificial neural networks, fuzzy system, graphs etc."* , and are not necessarily coupled to any evolutionary components at all, requiring simply a method for credit assignment. This enticing view offers the promise of memes capturing useful structural and behavioural patterns which can be carried between instances of the same problem, as is being explored in, for example, [431].

As this short selection of motivating considerations suggests, there are a number of diverse reasons why the hybridisation of evolutionary algorithms with other techniques is of interest to both the researcher and the practitioner. The use of other techniques and knowledge to augment the EA has been given various names in research papers such as: hybrid GAs, Baldwinian EAs, Lamarckian EAs, genetic local search algorithms, and others. Moscato [308] coined the name **memetic algorithm** (MA) to cover a wide range of techniques where evolutionary search is augmented by the addition of one or more phases of local search, or by the use of problem-specific information. The field is now sufficiently mature and distinct to have its own journal, annual workshop, and special issues of major journals dedicated to it.

10.2 A Brief Introduction to Local Search

In Section 3.7 we briefly described **local search** as an iterative process of examining the set of points in the neighbourhood of the current solution, and replacing it with a better neighbour if one exists. In this section we give a brief introduction to local search in the context of memetic algorithms. For more information there are a number of books on optimisation that cover local search in more detail, such as [3]. A local search algorithm can be illustrated by the pseudocode given in Fig. 10.2.

```
BEGIN
  /* given a starting solution i and a neighbourhood function n */
  set best = i;
  set iterations = 0;
  REPEAT UNTIL ( depth condition is satisfied ) DO
    set count = 0;
    REPEAT UNTIL ( pivot rule is satisfied ) DO
      generate the next neighbour j ∈ n(i);
      set count = count + 1;
      IF (f(j) is better than f(best)) THEN
        set best = j;
      FI
    OD
    set i = best;
    set iterations = iterations + 1;
  OD
END
```

Fig. 10.2. Pseudocode of a local search algorithm

There are three principal components that affect the workings of this local search algorithm.

The first is the choice of **pivot rule**, which can be **steepest ascent** or **greedy ascent** (also known as *first ascent*). In the former, the condition for terminating the inner loop is that the entire neighbourhood $n(i)$ has been searched, i.e., $count =| n(i) |$; whereas in the latter the termination condition is $((count =| n(i) |)\ or\ (best \neq i))$, i.e., it stops as soon as an improvement is found. In practice it is sometimes necessary to only consider a randomly drawn sample of size $N <<| n(i) |$ if the neighbourhood is too large to search.

The second component is the **depth** of the local search, i.e., the termination condition for the outer loop. This lies in the continuum between only one improving step being applied ($iterations = 1$) to the search continuing to local optimality: $((count =| n(i) |)\ and\ (best = i))$. Considerable attention has been paid to studying the effect of changing this parameter within MAs [211], and it can be shown to have an effect on the performance of the local search algorithm, both in terms of time taken, and in the quality of solution found.

The third, and most important, factor that affects the behaviour of the local search is the choice of neighbourhood generating function. In practice $n(i)$ is often defined in a operational way, that is, as a set of points that can be reached by the application of some move operator to the point i. An equivalent representation is as a graph $G = (v, e)$, where the set of vertices v are the points in the search space, and the edges relate to applications of the move operator i.e., $e_{ij} \in G \iff j \in n(i)$. The provision of a scalar fitness value f defined over the search space means that we can consider the graphs defined by different move operators as fitness landscapes [238]. Merz and Freisleben [293] present a number of statistical measures that can be used to characterise fitness landscapes, and that have been proposed by various authors as potential measures of problem difficulty. Merz and Freisleben show that the choice of move operator can have a dramatic effect on the efficiency and effectiveness of the local search, and hence of the resultant MA.

In some cases, domain-specific information may be used to guide the choice of neighbourhood structure within the local search algorithms. However, it has recently been shown that the optimal choice of operators can be not only instance specific within a class of problems [293, pp. 254–258], but also dependent on the state of the evolutionary search [264]. This result is not surprising when we consider that points that are locally optimal with respect to one neighbourhood structure may not be locally optimal with respect to another, unless of course they are globally optimal. Thus if a set of points has converged to the state where all are locally optimal with respect to the current neighbourhood operator, then changing the neighbourhood operator may provide a means of progression, in addition to recombination and mutation. This observation has also been applied in other fields of optimisation and forms the heart of methods such as the *variable neighbourhood search* algorithm [208] and *Hyperheuristics* [89, 246, 72, 71].

10.2.1 Lamarckianism and the Baldwin Effect

The framework of the local search algorithm outlined above works on the assumption that the current incumbent solution is always replaced by the fitter neighbour when found. Within a memetic algorithm, we can consider the local search stage to occur as an improvement or developmental learning phase within the evolutionary cycle, and (taking our cue from biology) we should consider whether the changes made to the individual (*acquired traits*) should be kept in the genotype, or whether the resulting improved fitness should be awarded to the original (pre-local search) member of the population.

The issue of whether acquired traits could be inherited by an individual's offspring was a major issue in the nineteenth century, with Lamarck arguing in favour. By contrast, the **Baldwin effect** [34] suggests a mechanism whereby evolutionary progress can be guided towards favourable adaptation without the changes in individuals' fitness arising from learning or development being reflected in changed genetic characteristics. Modern theories of genetics strongly favour the latter viewpoint. Pragmatically, we saw in Sect. 2.3.2 that the mapping from DNA to protein is highly complex and nonlinear, let alone the complexity of the developmental process by which the mature phenotype is created. In the light of this, it is hardly credible to believe that a process of reverse engineering could go on, coding the effects of phenotypically acquired traits back into the genotype.

Luckily, working within the medium of computer algorithms we are not restricted by these biological constraints, and so in practice both schemes are usually possible to implement within a memetic algorithm. In general, MAs are referred to as Lamarckian if the result of the local search stage replaces the individual in the population, and Baldwinian if the original member is kept, but has as its fitness the value belonging to the outcome of the local search process. In a classic early study, Hinton and Nowlan [215] showed that the Baldwin effect could be used to improve the evolution of artificial neural networks, and a number of researchers have studied the relative benefits of Baldwinian versus Lamarckian algorithms [224, 287, 435, 458, 459]. In practice, most recent work has tended to use either a pure Lamarckian approach, or a probabilistic combination of the two approaches, such that the improved fitness is always used, and the improved individual replaces the original with a given probability.

10.3 Structure of a Memetic Algorithm

There are a number of ways in which an EA can be used in conjunction with other operators and/or domain-specific knowledge as illustrated by Fig 10.3. A full taxonomy of possibilities can be found in [265].

10.3.1 Heuristic or Intelligent Initialisation

The most obvious way in which existing knowledge about the structure of a problem or potential solutions can be incorporated into an EA is in the initialisation phase. In our discussion of this issue in Sect. 3.5 we gave reasons why this might not be worth the effort in general, cf. Fig. 3.6. However, starting the EA by using existing solutions can offer interesting benefits:

1. It is possible to avoid reinventing the wheel by using existing solutions. Preventing waste of computational efforts can yield increased efficiency (speed).
2. A nonrandom initial population can direct the search into particular regions of the search space that contain good solutions. Biasing the search can result in increased effectivity (quality of end solution).

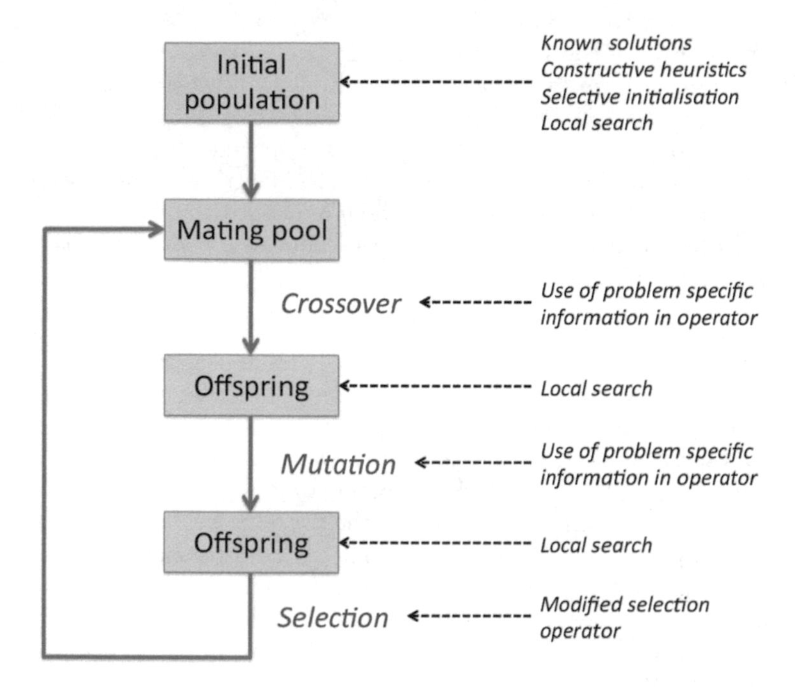

Fig. 10.3. Possible places to incorporate knowledge or other operators within the evolutionary cycle

3. All in all, a given total amount of computational effort divided over heuristic initialisation and evolutionary search might deliver better results than spending it all on 'pure' evolutionary search, or an equivalent multistart heuristic.

There are a number of possible ways in which the initialisation function can be changed from simple random creation, such as:

- **Seeding** the population with one or more previously known good solutions arising from other techniques. These techniques span the range from human trial and error to the use of highly specialised greedy constructive heuristics using instance-specific information. Examples of the latter include nearest-neighbour heuristics for TSP-like problems, 'schedule hardest first' for scheduling and planning problems, and a wealth of other techniques for different problems, which can be found in the operations research literature.
- In **selective initialisation** a large number of random solutions are created and then the initial population is selected from these. Bramlette [66] suggests that this should be done as a series of N k-way tournaments rather than by selecting the best N from $k \cdot N$ solutions. Other alternatives in-

clude selecting a set based not only on fitness but also on diversity so as to maximise the coverage of the search space.
- Performing a local search starting from each member of the initial population, so that the initial population consists of a set of points that are locally optimal with respect to some move operator.
- Using one or more of the above methods to identify one (or possibly more) good solutions, and then cloning them and applying mutation at a high rate (**mass mutation**) to produce a number of individuals in the vicinity of the start point.

All of these methods have been tried and have exhibited performance gains for certain problems. However, the important issue of providing the EA with sufficient diversity for evolution to occur must also be considered. In [421] Surry and Radcliffe examined the effect of varying the proportion of the initial population of a GA that was derived from known good solutions. They concluded that the use of a small proportion of derived solutions in the initial population aided genetic search, and as the proportion was increased, the *average* performance improved. However, the *best* performance came about from a more random initial population. In other words, as the proportion of solutions derived from heuristics used increased, so did the mean performance, but the variance in performance decreased. This meant that there were fewer really bad runs, but also fewer really good runs. For a certain type of problems (in particular, design problems as discussed in Chap. 9) this is an undesirable property.

10.3.2 Hybridisation Within Variation Operators: Intelligent Crossover and Mutation

A number of authors have proposed so-called intelligent variation operators, which incorporate problem- or instance-specific knowledge. At their most simple, these might take the form of introducing bias into the operators. To give a simple example, if a binary-coded GA is used to select features for use in another classification algorithm, one might attempt to bias the search towards more compact feature sets via the use of a greater probability for mutating from the allele value “use” to “don’t use” rather than vice versa. A related approach can be seen in [392], where genes encode for microprocessor instructions, which group naturally into sets with similar effects. The mutation operator was then biased to incorporate this expert knowledge, so that mutations were more likely to occur between instructions in the same set than between sets.

A slightly different example of the use of problem-specific (rather than instance-specific) knowledge can be seen in the modified one-point crossover operator used for protein structure prediction in [436]. Here the authors realised that the heritable features being combined by recombination were folds, or fragments of three-dimensional structure. A property of the problem is that during folding protein structures can be free to rotate about peptide bonds. The modified operator made good use of this knowledge by explicitly testing all the possible different orientations of the two fragments (accomplished by trying all the possible allele values in the gene at the crossover point), in order to find the most energetically favourable. If no feasible conformation was found, then a different crossover point was selected and the process repeated. This can be seen as a simple example of the incorporation of a local search phase into the recombination operator. Note that this should be distinguished from the simpler "crossover hill-climber" proposed in [238], in which all of the $l-1$ possible offspring arising from one-point crossover are constructed and the best chosen.

At the other end of the scale, at their most complex, the operators can be modified to incorporate highly specific heuristics, which make use of instance-specific knowledge. A good example of this is Merz and Freisleben's distance-preserving crossover (DPX) operator for the TSP [178]. This operator has two motivating principles: making use of instance-specific knowledge, while at the same time preserving diversity within the population to prevent premature convergence. Diversity is maintained by ensuring that the offspring inherits all of the edges common to both parents, but none of the edges that are present in only one parent, and is thus at the same distance to each parent as they are to each other. The intelligent part of the operator comes from the use of a nearest-neighbour heuristic to join together the subtours inherited from the parents, thus explicitly exploiting instance-specific edge length information. It is easy to see how this type of scheme could be adapted to other problems, via the use of suitable heuristics for completing the partial solutions after inheritance of the common factors from both parents.

10.3.3 Local Search Acting on the Output from Variation Operators

The most common use of hybridisation within EAs, and that which fits best with Dawkins' concept of the meme, is via the application of one or more phases of improvement to individual members of the population during the EA cycle, i.e., local search acting on whole solutions created by mutation or recombination. As is suggested from Fig. 10.3, this can occur in different places in cycle i.e., before or after selection or after crossover and/or mutation, but a typical implementation might take the form given in Fig. 10.4

The natural analogies between human evolution and learning, and EAs and artificial neural networks (ANNs) prompted a great deal of research into the use of EAs to evolve the structure of ANNs, which were then trained using back-propagation or similar means during the 1980s and early 1990s. This research gave a great deal of insight into the role of learning, Lamarckianism, and the Baldwin effect to guide evolution (e.g., [215, 224, 287, 435, 458, 459] amongst many others), and served to reinforce messages that were proposed

```
BEGIN
  INITIALISE population;
  EVALUATE each candidate;
  REPEAT UNTIL ( TERMINATION CONDITION is satisfied ) DO
    SELECT parents;
    RECOMBINE to produce offspring;
    MUTATE offspring;
    EVALUATE offspring;
    [optional] CHOOSE Local Search method to apply;
    IMPROVE offspring via Local Search;
    [optional] Assign Credit to Local Search methods;
    (according to fitness improvements they cause);
    SELECT individuals for next generation;
    REWARD currently successful Local Search methods;
    (by increasing probability that they are used);
  OD
END
```

Fig. 10.4. Pseudocode for a simple memetic algorithm with optional choices for multiple memes

by "real-world" practitioners for several years as to the usefulness of incorporating local search and domain-based heuristics. Since then a number of PhD theses [211, 259, 267, 292, 309] have provided the beginnings of a theoretical analysis, and both theoretical and empirical results to justify an increased interest in these algorithms.

One recent result of particular interest to the practitioner is Krasnogor's formal proof that, in order to reduce the worst-case run times, it is necessary to choose a local search method whose move operator is not the same as those of the recombination and mutation operators [259]. This formalises the intuitive point that within an MA recombination, and particularly mutation, have valuable roles in generating points that lie in different basins of attraction with respect to the local search operator. This diversification is best done either by an aggressive mutation rate, or preferably by the use of a variation operators that have different neighbourhood structures.

10.3.4 Hybridisation During Genotype to Phenotype Mapping

A widely used hybridisation of memetic algorithms with other heuristics is during the genotype–phenotype mapping prior to evaluation. A good example of this is the use of instance-specific knowledge within a decoder or a repair function, as seen in Sect. 3.4.2, where we can consider the decoder function for the knapsack problem as being a packing algorithm that takes its inputs in the order suggested by the EA.

This approach, where the EA is used to provide the inputs controlling the application of another heuristic, is frequently used to great effect in, for example, timetabling and scheduling problems [210], and in the "sector first–order second" approach to the vehicle routing problem [428].

As can be seen, there is a common thread to all of these approaches, which is to make use of existing heuristics and domain information wherever possible. The role of the EA is often that of enabling a less biased application of the heuristics, or of problem decomposition, so as to permit the use of sophisticated, but badly scaling heuristics when the overall problem size would preclude their use.

10.4 Adaptive Memetic Algorithms

Probably the most important factor in the design of a memetic algorithm incorporating local search or heuristic improvement is the choice of improving heuristic or local search move operator, that is to say, the way that the set of neighbouring points — to be examined when looking for an improved solution — is generated.

To this end, a large body of theoretical and empirical analysis of the utility of various statistical measures of landscapes for predicting problem difficulty is available [239]. Merz and Freisleben [293] consider a number of these measures in the context of memetic algorithms, and show that the choice of move operator can have a dramatic effect on the efficiency and effectiveness of the local search, and hence of the resultant MA. We have already mentioned Krasnogor's PLS complexity analysis result, which suggests that to reduce the worst-case time complexity of the algorithm it is desirable for the move operator of the LS to define a different landscape to the mutation and crossover operators.

In general then, it is worth giving careful consideration to the choice of move operators used when designing a MA: for example, using 2-opt for a TSP problem might yield better improvement if not used in conjunction with the inversion mutation operator described in Sect. 4.5.1. In some cases, domain-specific information may be used to guide the choice of neighbourhood structure within the local search algorithms.

One simple way to surmount these problems is the use of *multiple* local search operators in tandem, in a similar fashion to the use of multiple variation operators seen in Chapter 8. Krasnogor and Smith [264] introduced the idea of what they first called 'multimeme' algorithms, where the evolutionary algorithm was coupled with not one, but several, local search methods, together with some mechanism for choosing between them according to their perceived usefulness at any given stage of search. In this case they used self-adaptation, so that each candidate solution carried a gene which indicated which local search method to use. This was inherited from its parents and was subject to mutation, in much the same way that Self-adaptation is widely used to adapt the choice of mutation rates (see, for example, Sect. 4.4.2). An example of this can be seen in [261],where a range of problem specific move operators, such as local stretches, rotations, and reflections, each tailored to different stages of the folding process, are used for a protein structure prediction problem.

Working with continuous representations, Ong and Keane applied similar ideas, but used a different choice mechanism [326] in what they called 'Meta-Lamarckian Learning'. The commonality between these approaches was quickly recognised, and [328] present an excellent recent review of work in the field of what they term "Adaptive Memetic Algorithms". This encompasses Multi-memetic Algorithms ([258, 259, 263, 264, 261]), the coevolving memetic algorithms (COMA) framework ([384, 387, 388, 389, 381]), Meta-Lamarckian MAs ([326]), Hyperheuristics ([89, 246, 72, 71]), and Self-generating MAs ([260, 262]).

Essentially, all of these approaches maintain a pool of local search operators available to be used by the algorithm, and at each decision point make a choice of which to apply. Ong's classification uses terminology developed elsewhere to describe adaptation of operators and parameters in evolutionary algorithms (see Chap. 8). This taxonomy categorises algorithms according to the way that these decisions are made. One way ('static') is to use a fixed strategy. Another ('adaptive') uses feedback of which operators have provided the best improvement recently, and is further subdivided into "external", "local" (to a deme or region of search space), and "global" (to the population) according to the nature of the knowledge considered. Finally, they note that LS operators may be linked to candidate solutions ('self-adaptive'). As with the field of parameter tuning and control, a lot of research has been focussed on good ways of *identifying* currently helpful local search mechanisms (also known as credit assignment [381, 391]), *rewarding* useful memes by increasing their probability of being used, and *adapting* the definitions of the local search mechanisms.

Based on these ideas, Meuth et al. [294] distinguished between:

- First-Generation MAs — which they define as "Global search paired with local search",
- Second-Generation MAs — "Global search with multiple local optimizers. Memetic information (Choice of optimizer) passed to offspring (Lamarckian evolution)",
- Third-Generation MAs: — "Global search with multiple local optimizers. Memetic information (choice of local optimizer) passed to offspring (Lamarckian evolution). A mapping between evolutionary trajectory and choice of local optimizer is learned".

They noted that at the time of writing the Self-generating MAs, and COMA are the only algorithms falling into the 3G class, and go on to propose (but not implement) a fourth generation of MAs in which they suggest: "Mechanisms of recognition, generalization, optimization, and memory are utilized". Arguably the use of pattern-based memes in COMA falls into this class, and certainly the framework described in [82] represents an important step towards such algorithms.

A number of subsequent papers have expanded on these themes. Barkat-Ullah et al. [40] propose an agent-based approach for optimising constrained problems defined over continuous spaces. Here each agent has available to it a suite of local search algorithms, and maintains a *local* record for each. This is a scalar value in the range $\{-1,1\}$ according to the meme's effect on the feasibility of the candidate solution, and also on the fitness improvement caused in the last generation. Nguyen et al. [319] proposed static adaption in cellular memetic algorithms. Their approach split the population into groups according to fitness diversity and applied global search to one member from each group, with a blacklist of members that did not benefit from local search. This blacklist method uses *local* historical evidence to bias the global / local search tradeoff, and while highly effective, it will of course only work with fixed memes. A more general probabilistic memetic framework was proposed in [320] to adapt the global/local search trade-off. Using arguments based on the likelihood of generating points within a basin of attraction on a continuous landscape, they proposed dynamically estimating the probabilities of achieving benefits via local search and global search, and adapting the number of iterations allowed to local search accordingly. This was successfully instantiated using *local* search traces plus a database of historical points.

What is notable about many of these advanced algorithms is that although they avoid some of the issues such as robustness, which arise from a single fixed choice of meme, their focus highlights some of the common design issues that are faced when implementing an MA, which we now turn to.

10.5 Design Issues for Memetic Algorithms

So far we have discussed the rationale for the use of problem-specific knowlege or heuristics within EAs, and some possible ways in which this can be done. However, as ever, we must accept the caveat that, like any other technique, MAs are not some 'magic solution' to optimisation problems, and care must be taken in their implementation. In the sections below we briefly discuss some of the issues that have arisen from experience and theoretical reasoning.

Preservation of Diversity

The problem of premature convergence, whereby the population converges around some suboptimal point, is recognised within EAs but is exacerbated in MAs by the effect of local search. If the local search phase continues until each point has been moved to a local optimum, then this leads to an inevitable loss of diversity within the population.[3] A number of approaches have been developed to combat this problem such as:

- When initialising the population with known good individuals, only using a relatively small proportion of them.
- Using recombination operators which are designed to preserve diversity.
- Modifying the selection operator to prevent duplicates.
- Modifying the selection operator or local search acceptance criteria to use a Boltzmann method so as to preserve diversity.

This last method bears natural analogies to simulated annealing [2, 250], where worsening moves can be accepted with nonzero probability, to aid escape from local optima, see also Sect. 8.4.5. A promising method that tackles the diversity issue explicitly is proposed in [263], where during the local search phase a less-fit neighbour may be accepted with a probability that increases exponentially as the range of fitness values in the population decreases:

$$P(accept) = \begin{cases} 1 & \text{if } \Delta E > 0, \\ e^{\frac{k \Delta E}{F_{max} - F_{avg}}}, & \text{otherwise,} \end{cases}$$

where k is a normalisation constant and we assume a maximisation problem, $\Delta E = F_{\text{neighbour}} - F_{\text{original}}$.

Recently this issue has been revisited in the context of adaptive MAs, using insights from the behaviour of different local search mechanisms. Neri and Caponio [77] propose a 'Fast Adaptive Memetic Algorithm' which simultaneously adapts the global and local search characteristics according to a measure of (*global*) fitness diversity. The former is done by adjusting the EA's population size and the 'aggressiveness' of mutation to maintain diversity. The probability of applying two very different local search operators is determined using a static external rule, taking as evidence the generation count and the ratio of the current fitness diversity to the *extremal* value ever observed. This concept is explored further in [316, 317].

Use of Knowledge

A final point that might be taken into consideration when designing a new memetic algorithm concerns the use and reuse of knowledge gained during the optimisation process. To a certain extent this is done automatically by recombination, but, generally speaking, explicit mechanisms are not used.

One possible hybridisation that explicitly uses knowledge about points already searched to guide optimisation is with **tabu search** [185]. In this algorithm a "tabu" list of visited points is maintained, to which the algorithm is forbidden to return. Such methods appear to offer promise for maintaining

[3] Apart from the exceptional case where each member of the population lies within the basin of attraction of a different local optimum.

diversity. Similarly, it is easy to imagine extensions to the Boltzmann acceptance/selection schemes that utilise information about the spread of genotypes in the current population, or even past populations, when deciding whether to accept new solutions.

10.6 Example Application: Multistage Memetic Timetabling

In order to illustrate some of the ways that EAs can be combined with other techniques, we take as our example an application to examination timetabling described in [73]. Timetabling is, in general, an NP-complete problem, and the examination timetabling application is particularly beloved of academic researchers, not least because they are regularly made aware of its importance and difficulty. The general form that the problem takes is of a set of examinations, E, each of which has a number of seats required, s_i, to schedule over a set of time periods P. Usually a co-occurrence matrix C is provided, where c_{ij} gives the number of students sitting both exams i and j. If feasible solutions exist, then this is a constraint satisfaction problem, but in general this might not be the case, so it is more common to take an indirect approach and treat it as a constrained optimisation problem via the use of a penalty function.[4] This function considers a number of terms

- Exams can only be scheduled in rooms with adequate capacity.
- It is highly undesirable to timetable two exams i and j at the same time if $c_{ij} > 0$, since this requires quarantining those students until they can sit the second paper.
- It is not desirable to have students sitting two exams on the same day.
- It is preferable not to have students sitting exams in consecutive periods, even if there is a night between them.

This timetabling problem has been well studied, and many heuristic approaches have been proposed, but a common problem has been that they do not always scale well. The approach documented by Burke and Newell is particularly interesting and is relevant to this chapter because it has the following features:

- A decomposition approach is taken, whereby a heuristic scheduler breaks down the problem into smaller parts that can be more easily solved by an optimisation technique.
- The optimisation heuristic used is an EA.
- The EA itself incorporates other heuristics, i.e., it is an MA in its own right.

[4] See Chap. 13.

The heuristic scheduler divides the set E into a number of equal-sized smaller sub-groups, which are scheduled in turn. Thus when scheduling the elements of the nth subgroup, the elements of the previous $n - 1$ subgroups are already in place and cannot be altered. The set E is partitioned by using a metric to estimate the difficulty of scheduling each exam, and then ranking the exams so that the most difficult ones are scheduled first. Three different metrics are considered, based on the number of conflicts with other events, the number of conflicts with *previously scheduled* events, and the number of valid periods for an event left in the timetable. The authors also consider the use of look-ahead techniques in which two subsets are considered, but only one timetabled. This strategy could, of course, be used with any one of a number of techniques embedded within it to handle the timetabling of each subset in turn. The heuristic chosen is itself a memetic algorithm with the parameters listed in Table 10.1.

Representation	Set of linked lists of exams, each encoding for one period
Recombination	None
Mutation	Random choice of "light" or "heavy" mutation
Mutation probability	100%
Parent selection	Exponential ranking
Survival selection	Best 50 of 100 offspring
Population size	50
Initialisation	Randomly generated then local search applied to each
Termination condition	Five generations with no improvement in best fitness
Special features	Local search (to local optimum) applied after mutation

Table 10.1. Table describing MA embedded with multistage timetabling algorithm

As can be seen, there are several points worth commenting on. Each member of the initial population is created by generating a random permutation of the exams, and then (in that order) assigning each one to the first valid period. The local search algorithm is always applied until a locally optimal solution is reached, but it uses a greedy ascent mechanism so there is some variability in the output. It is applied to each initial solution, and to each offspring, thus the EA is always working with solutions that are locally optimal, at least with respect to this operator.

The authors reported that previous experiments motivated them against the use of recombination for this problem, but instead each offspring is created via the use of one of two problem-specific mutation operators. The "light" operator is a version of the scramble mutation that checks for the feasibility of the solutions it produces. The "heavy" mutation operator is highly instance specific. It looks at the parent and calculates a probability of "disrupting" the events in each period based on the amount of penalty it seems to be causing. However, this operator also makes use of knowledge from other solutions, since these probabilities of disruption are modified by reference to the best-performing solution in the current population.

The results obtained by this algorithm are impressive, both in terms of speed and quality of solutions. It is worth emphasizing the following points that have led to this success:

- The combination of heuristic sequencing with an EA-based approach finds better results faster than either approach does on its own.
- The algorithm uses local search so that its initial population is already considerably better than random.
- Strong selection pressure is applied: both exponential ranking for parent selection plus (50,100) survivor selection.
- Intelligent mutation operators are used. One uses instance-specific information to prevent it from producing solutions that violate the most important constraints. The second is highly problem specific, aimed at disrupting 'poor' periods.
- The "heavy" mutation operator makes use of information from the rest of the population to decide how likely it is to disrupt a period.
- The depth of local search is always maximal, i.e., the parent population will only come from the set of local optima of the local search landscape.
- Despite the strong selection pressure, and the point above, the fact that mutation is always applied, and that all the search operators have different underlying move mechanisms, means that a premature loss of diversity is avoided.
- As detailed in the paper, there is major use of a variety of coding and algorithmic strategies to avoid full evaluation of solutions and to speed up the manipulation of partial solutions.

For exercises and recommended reading for this chapter, please visit `www.evolutionarycomputation.org`.

11

Nonstationary and Noisy Function Optimisation

Unlike most of the examples we have used so far, real-world environments typically contain sources of uncertainty. This means that if we measure the fitness of a solution more than once, we will not always get the same result. Of course, biological evolution happens in just such a dynamic environment, but there are also many EA applications in environments featuring change or noise when solutions are evaluated. In these **nonstationary** situations the search algorithm has to be designed so that it can compensate for the unpredictable environment by monitoring its performance and altering some aspects of its behaviour. An objective of the resulting adaptation is not to find a single optimum, but rather to select a sequence of values over time that maximise or minimise some measure of the evaluations, such as the average or worst. This chapter discusses the various sources of unpredictability, and describes the principal adaptations to the basic EA in response to them.

11.1 Characterisation of Nonstationary Problems

At this stage we must consider some basic facts about the process of going from a representation of a solution (genotype) x, to measuring the quality of a candidate solution for the task at hand, $f(x)$. For illustration we will use a simple household example — designing a mop for cleaning spills of various different liquids from a floor. We will assume that the candidate solution in fact describes the structure of the sponge — that is to say the size of pores, elasticity, shape of contact area, etc.[1] The quality recorded for a given solution may be unpredictable for one or more of the following reasons.

[1] In general, δq stands for a small random change in the value of some property q.

The Genotype to Phenotype Mapping Is Not Exact and One-to-One

If the fitness is to be measured via simulations, the genotype may use double-precision floating point numbers to code for the design parameters, but in the simulation there might be differences in the resolution of the models. If the fitness is to be measured via physical effects, the manufactured phenotype may not perfectly reflect the encoded design parameters. Thus, for our example, we might be measuring the cleaning ability of a slightly different sponge to the one we thought. In terms of a search landscape, the fitness observed may be that of a point in the region of x: $f_{observed}(x) = f(x + \delta x)$.

The Act of Measurement Itself Is Prone to Error or Uncertainty

This might arise from, for example, human error, small random fluctuations in shape of physical objects as their molecules vibrate, randomness in the movement of electrons through a sensor or wire, or the collected randomness of more complex organisms such as people, markets, packets on computer network or physical traffic. In terms of our mop, we might mismeasure the quantities of fluid absorbed by the sponge. This means a rethink of our landscape metaphor: the unique surface in the altitude dimension representing the fitness of points in the search space is replaced by a 'cloud' — a probability distribution from which we sample when we measure fitness, whose 'thickness' may vary across the space. Many different models may be used to characterise the noise. The most straightforward is to break down a quality function into two components: $f_{observed}(x) = f_{mean}(x) + f_{noise}(x)$. Here the first component represents the average that we would find if we measured fitness many times, and the second noise component is typically modelled as a random drawing from a normal distribution $N(0, \sigma)$.

The Environment Changes Over Time

This may be because the external environment is inherently volatile, or it may be that the very act of evaluating solutions affects subsequent fitness. For example, in an interactive EA, each interaction potentially increases user fatigue and changes the expectations (Sect. 14.1). If our mop is being tested in an environment with significant seasonal fluctuations in temperature, then this may affect either the absorbency of the sponge material, or the viscosity of fluids tested. This could mean that if we measured the same design every day for a year we would observe seasonal cyclic changes in the fitness. In the context of a search landscape, this means that the locations of the optima are now time-dependent, i.e., $f_{observed}(x) = f(x, t)$.

In many real-world problems, one or more of these effects occur in combination. It remains for the algorithm designer to decide which will be present, take account of the context in which the tool created will be used, and select appropriate modifications from those listed in subsequent sections.

11.2 The Effect of Different Sources of Uncertainty

A number of researchers have proposed mechanisms for dealing with uncertainty, and examined their performance on test cases and real-world problems. Algorithms are typically compared by running them for a fixed period and calculating two time-averaged metrics, which correspond to different types of real-world applications.

The first of these is the online measure [102] and is simply the average of all calls to the evaluation function during the run of the algorithm. This measure relates to applications where it is desirable to maintain consistently good solutions, e.g., online process control [164, 444] or financial trading. The second metric considered is the offline performance and is the time-averaged value of the best-performing member of the current population. Unlike the online metric, offline performance is unaffected by the occasional generation of individuals with very low fitness, and so is more suitable for problems where the testing of such individuals is not penalised, e.g., parameter optimisation using a changing design model.

If we use the time-dependent notation for the fitness function as $f(x,t)$ and denote the best individual in a population $P(t)$ at time t by $best(P(t))$, then we can formalise the two metrics over a period T as follows:

$$online = \frac{1}{T} \times \sum_{t=1}^{T} \frac{1}{|P(t)|} \sum_{x \in P(t)} f(x,t)),$$

$$offline = \frac{1}{T} \times \sum_{t=1}^{T} f((best(P(t)),t)).$$

Finally, let us note that in some cases it may be appropriate to consider both metrics in a multiobjective approach, since optimising the mean fitness may be the principle desiderata, but evaluating low-fitness solutions might be catastrophic. In this case one approach might be to use a surrogate model to screen out such potential fatal errors.

The three different sources of uncertainty identified in the previous section affect the performance of the EA in different ways. Considering errors in the genotype–phenotype mapping, we can note that using the average of n repeated fitness evaluations $\frac{1}{n} \times \sum_n f(x + \delta x)$ to determine the fitness of any given x means using a sample of n points from the neighbourhood around x. However, the sampled neighbourhoods of adjacent solutions can overlap, that is, $x + \delta x$ can coincide with $y + \delta y$. Hence, fine-grained features of the fitness landscape will be smoothened out possibly removing local optima in the process. In practice this is often a good thing. From the search perspective it creates gradients around steps and plateaus in the landscape, and from a problem-solving perspective it reduces the attractiveness of high-quality solutions that are surrounded by much lower quality neighbours, which might be considered very ‘brittle’.

Considering noise in the act of measurement itself, the average of n repeated measurements is $f(x) + \frac{1}{n} \times \sum_n N(0, \sigma)$ and the second term will approximate zero as n is increased. In other words, when repeatedly sampling from random noise, the deviations cancel out and you get an estimate of the mean. Unlike the case above, there is no smoothing of landscape features. This is illustrated in Figure 11.1, which shows a fitness function $f(x) = 1/(0.1 + x^2)$ and the values estimated after five samples with two different sorts of uncertainty present. Both types of noise were drawn uniformly from a distribution between $+/-$ 0.4. Already becoming apparent is that the errors in the genotype–phenotype mapping reduce the height of the estimated local optimum and make it wider. In contrast, the effect of noise in the measurement alone is already being reduced to near-zero after five samples.

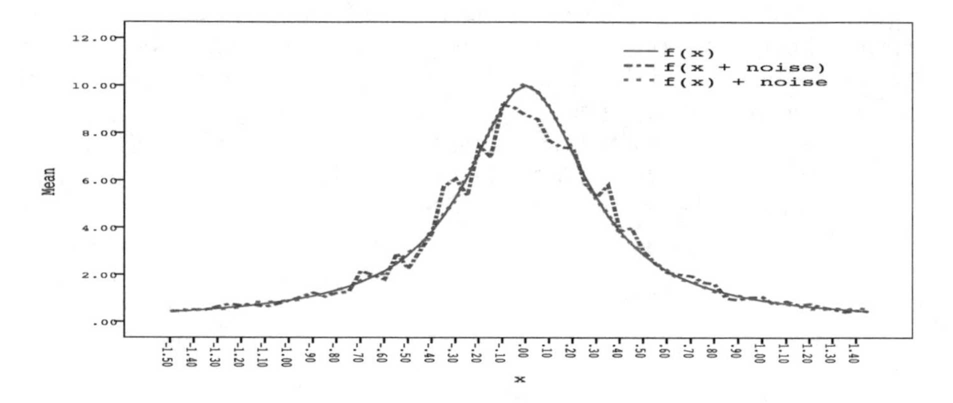

Fig. 11.1. Effect of different types of uncertainty on estimated fitness. Curves show mean values estimated after five samples for each value of x

Regarding the third situation of nonstationary fitness functions, Cobb [84] defines two ways of classifying these:

- Switching versus continuous, which is based on time-scale of change with respect to the rate of evaluation — the former providing sudden and the latter more gradual changes. Continuous changes might be cyclical (e.g., related to seasonal effects) or reflect a more uniform movement of landscape features (for example, when arising from gradual wear-and-tear of physical parts).
- Markovian versus state dependent. In the first of these, the environment at the next time step is purely derived from the current one, whereas in the latter far more complex dynamics may play out.

To illustrate these differences, consider a simple example many of us encounter daily – modelling and predicting traffic levels on a commute. These will change gradually during each day, building up to, and then tailing off from, peak values at rush hours. When planning a journey we would expect that the likely travel time of a direct route involving a busy road will vary during the day. However, there will also be more switching behaviour as the flows are affected by one-off events such as public holidays, and between different periods — for example the amount of variation and overall levels of traffic are often less during school holidays. On these days we might happily decide take the most direct route. Considering ten-minute intervals over a single day in a city, we might view traffic levels as driven by some Markovian process — the value at the next period depends on traffic levels now, but not prior to that, since the cause is the aggregated behaviour of lots of independent people, mostly travelling to, from, and in the course of work. However, if we turn our attention to modelling air traffic, the situation is still clearly time-varying, but is highly state-dependent as disruption to an airline's schedule in one place can have enormous knock-on effects due to planes being in the wrong place, etc.

11.3 Algorithmic Approaches

11.3.1 Approaches That Increase Robustness or Reduce Noise

The only viable approach for reducing the effect of noise, whether in the fitness function evaluation, or in the genotype-to-phenotype mapping, is to repeatedly re-evaluate solutions and take an average. This might be done either explicitly, or (as described further below) implicitly via the population management processes of parent and survivor selection.

The principle question that arises in explicit approaches is, how many times should the fitness be sampled? Bearing in mind that EAs naturally contain some randomness in their processes anyway, the key issue from the perspective of evolution is being able to reliably distinguish between good and bad members of the population. Thus, a common approach is to monitor the degree of variation present, and resample when this is greater than the range of estimated fitnesses in the population. This reasoning suggests that the rate of resampling would increase as the population converges towards high-quality solutions.

When calculating how large a sample to take, there is also the law of diminishing returns: in general, the standard deviation observed decreases only as fast as the square root of the number of measurements taken.

Finally, it is worth mentioning that it is often worth the extra book-keeping of making resampling decisions independently for each solution, since the amount of noise will often not be uniform across the search space.

11.3.2 Pure Evolutionary Approaches to Dynamic Environments

The distributed nature of the genetic search provides a natural source of power for exploring in changing environments. As long as sufficient diversity remains in the population, the EA can respond to a changing search landscape by reallocating future trials. However, the tendency of EAs, especially of GAs, to converge rapidly results in the population becoming homogeneous, which reduces the ability of the EA to identify regions of the search space that might become more attractive as the environment changes. In such cases it is necessary to complement the standard EA with a mechanism for maintaining a healthy exploration of the search space. (Recall the self-adaptation example from Sect. 4.4.2.)

In [84] the behaviour of a standard GA on a parabolic function with the optima moving sinusoidally in space was observed. This was done for a range of bitwise mutation rates. It was found that the offline performance decreased as the rate of change increased, for all mutation probabilities. As the rate of change increased, the mutation rate that gave optimal offline performance increased. Finally, it was noted that as problem difficulty increased, the rate of change that GA could track decreased.

In the light of these findings, various approaches have been proposed that are aimed at responding to different types of environmental change.

11.3.3 Memory-Based Approaches for Switching or Cyclic Environments

The first strategy expands the memory of the EA in order to build up a repertoire of ready responses for various environmental conditions. The main examples of this approach are the GA with diploid representation [194] and the structured GA [94]. Goldberg and Smith examined the use of diploid representation and dominance operators to improve performance of an EA in an oscillating environment [402], while Dasgupta and McGregor presented a modified "structured GA" with a multilayered structure of the chromosome which constitutes a "long-term distributed memory".

11.3.4 Explicitly Increasing Diversity in Dynamic Environments

The second modification strategy effectively increases diversity in the population directly (i.e., without extending the EA memory) in order to compensate for changes encountered in the environment. Examples of this strategy involve the GA with a hypermutation operator [84, 85], the random immigrants GA [199], the GA with a variable local search (VLS) operator [443, 444], and the thermodynamic GA [306].

The hypermutation operator temporarily increases the mutation rate to a high value, called the hypermutation rate, during periods when the time-averaged best performance of the EA worsens. In his 1992 study, Grefenstette noted that under certain conditions hypermutation might never get triggered [199].

The random immigrants mechanism replaces a fraction of a standard GA's population by randomly generated individuals in each generation in order to maintain a continuous level of exploration of the search space. It was reported that 30% replacement gave the best off-line tracking: if the value is too high the algorithm is unable to converge between changes; however, off-line performance decreases with proportion replaced.

In an extensive study, Cobb and Grefenstette compared hypermutation with random immigrants and simple GA (with high mutation rate) [85]. They noted that there was a qualitative difference in the nature of the mutation operator in the three algorithms:

- Simple Genetic Algorithm (SGA) – uniform in population and time
- Hypermutation – uniform in population, not in time
- Random immigrants – uniform in time, not in population

They used two landscapes, and three types of change: a linear motion in the first problem (moving 1 step along an axis every 2 or 5 generations), randomly shifting the optima in the first problem every 20 generations, and swapping between the two problems every 2 or 20 generations. Their findings were:

- SGA: A high mutation probability of 0.1 was reasonably good at the translation tasks, but gave very poor online performance. It was unable to track the steadily moving optimum or oscillation. In general, the mutation probability needs to be matched to the degree of change.
- Hypermutation: High variances in performance were noted, and the higher mutation rate needed careful tuning to the problem instance. It was much better at tracking sudden changes than SGA and gave better online performance than SGA or random immigrants when the rate of change was slow enough to allow a lower rate of mutation.
- Random Immigrants: This strategy was not very good at tracking linear movement, but was the best at the oscillating task. They hypothesised that this was because it allowed the preservation of niches. The strategy displayed poor performance on stationary and slowly changing problems.

The VLS operator uses a similar triggering mechanism to hypermutation, and it enables local search around the location of the population members before the environmental change. The range of the search is gradually extended using a heuristic that attempts to match the degree of change.

The thermodynamic GA can maintain a given level of diversity in population by evaluating the entropy and free energy of the GA's population. The free energy function is effectively used to control selection pressure during the process of creating a new population.

11.3.5 Preserving Diversity and Resampling: Modifying Selection and Replacement Policies

In [441, 442] the suitability of generational GAs (GGAs) and steady-state GAs (SSGAs) was studied for use in dynamic environments. Results showed that the SSGA with a "delete-oldest" replacement strategy can adapt to environmental changes with reduced degradation of offline and particularly online performance. The improved performance of the SSGA can be explained by the fact that an offspring is immediately used as a part of the mating pool, making a shift towards the optimal solution possible in a relatively early phase of search. The authors concluded that the steady-state model was better suited tor use in nonstationary environments, particularly for on-line applications.

Selection is a vital force in any evolutionary algorithm, and an understanding of the nature of its effects is necessary if effective algorithms are to be developed. For GGAs selection has been well studied, and methods have been developed that reduce much of the noise inherent in the stochastic algorithm, e.g., SUS [32]. Unfortunately, the very nature of SSGAs precludes the use of such methods and those available are inherently more noisy.

In [400] a Markov chain analysis of the takeover probability versus time was used to investigate sources of noise in several replacement strategies. Some variations in performance arise from losing the only copy of the current best in the population, which happened approximately 50% of the time for delete random, and 10% of the time for delete-oldest. Performance comparisons on static landscapes demonstrated that the extent to which this affects the quality of the solutions obtained depends on the ability of the reproductive operators to rediscover the lost points. In [78] other strategies, e.g., deletion by exponential ranking, were also shown to lose the optimum.

A common way of avoiding this problem is to incorporate elitism, often in the form of a delete-worst strategy. Chakraborty [78, 79] showed that this exhibits increased selection pressure, which can lead to premature convergence and poor performance on higher dimensional problems.

In [399] a number of replacement strategies were compared in combination with two different ways of achieving elitism. The first was the common method described in Section 5.3.2, and the elite member can either be preserved with its original fitness value, or be reevaluated and the new fitness value saved. The second, "conservative selection" is an implicit mechanism introduced in [444]. Here each parent was selected by a binary tournament between a randomly selected member of the population and the member about to be replaced. If the latter is the current best, then it will win both tournaments, so recombination will have no effect, and (apart from the effects of mutation) elitism is attained. In [400] this was shown to guarantee takeover by the optimal class, but at a much slower rate than delete-worst or elitist delete-oldest. In total, ten selection strategies were evaluated for their online and offline performance on two different test problems. Deletion of the oldest, worst, and random members was done in conjunction with both standard and conservative tournaments. Additionally, a delete-oldest policy was tested with four variants of elitism. These were:

1. Oldest is kept if it is one of the current best, but is re-evaluated.
2. Oldest is kept if it is the sole copy of the current best and is re-evaluated.

3. As 1, but without re-evaluation (original fitness value kept).
4. As 2, but without re-evaluation (original fitness value kept).

This was done for algorithms with and without hypermutation on two different classes of problems. The results obtained clearly confirmed that for some algorithms an extra method for creating diversity (in this case, hypermutation) can improve tracking performance, although not all of the strategies tested were able to take advantage of this. However, two factors are immediately apparent from these results which hold with or without hypermutation.

Exploitation: Strategies such as delete-oldest or delete-random, which can lose the sole copy of the current population best, performed poorly. This matched the theoretical analysis and results on static landscapes noted above. Therefore some form of elitism is desirable.

Reevaluation: In potentially dynamic environments it is essential that the fitness of points on the landscape is continuously and systematically reevaluated. Failure to do so leads to two effects. First, the population can get 'dragged back' to the original peak position, as solutions near there are selected to be parents on the basis of out-of-date information. Second, it can also lead to a failure to trigger the hypermutation mechanism. Although this was obvious for the third and fourth variants of elitism tested, it also applies to the much more common delete-worst policy. In this case if the population had converged close to the optimum prior to the change, the worst members that get deleted may be the only ones with a true fitness value attached. The importance of systematic reevaluation was clear from the difference between conservative delete-oldest and conservative delete-random. The former always produced better performance than the latter, and very significantly so when hypermutation was present.

Of all the policies tested, the conservative delete-oldest was the best suited to the points noted above and produced the best performance. The improvement over the elitist policy with reevaluation is believed to result not merely from the reduced selection pressure, but from the fact that the exploitation of good individuals is not limited to preserving the very best, but will also apply (with decreasing probability) to the second-best member, and so on. Since the implicit elitism still allows changes via mutation, there is a higher probability of local search around individuals of high fitness, while worse members are less likely to win tournaments, and so they are replaced with offspring created by recombination. The result is that even without hypermutation the algorithm was able to track environmental changes of modest size.

11.3.6 Example Application: Time-Varying Knapsack Problem

This problem is a variant of that described in [306]. As discussed in Sect. 3.4.2, we have a number of items each having a value (v_i^t) and a weight or cost (c_i^t) associated with them, and the problem is to select a subset that maximises the sum of the elements' values while meeting a (time-varying) total capacity constraint $C(t)$.

In [399], Smith and Vavak outline a series of experiments on this problem aimed at investigating the effect of different survivor selection policies. In the particular case investigated, the values v_i and costs c_i attached to the items remained constant, but the capacity constraint $C(t)$ alternated between 50%, 30%, and 80% of C_{sum}, changing once every 20,000 evaluations.

The algorithm used was a binary-coded SSGA with 100 members. Parent selection was by binary tournaments, with the fitter member always selected. In some cases the conservative tournament selection operator was used. Uniform crossover was used (with probability 1.0) to generate offspring, as this shows no positional bias (Sect. 16.1). The rest of the parameter settings were decided after some initial experimentation to establish robust values.

The hypermutation operator was implemented as it is currently the most commonly used method for tracking. It was triggered if the running average of the best performing members of the population over an equivalent of three generations of the generational GA (in this case, 300 evaluations) drops by an amount that exceeds a predefined threshold. In this case a value of threshold TH=3 was used. The best performing member of the population was re-evaluated for 100 evaluations. Once it had been triggered, the hypermutation rate (0.2) was switched back to the baseline mutation rate (0.001) as soon as the best performing member of the population reached 80% of its value before the environmental change occurred. The setting of the parameters (80% and hypermutation rate 0.2) was found to provide good results for the given problem. A prolonged period of high mutation for values higher than 80% has a negative effect on on-line performance because diversity is introduced into the population despite the correct region of the search space having already been identified. Similarly to the choice of the threshold level described previously, the values of both parameters were selected empirically.

As hinted above, the best results came from the combination of conservative tournaments for parent selection policy, with a delete-oldest policy. Here each member has a fixed lifespan, but when its turn comes to be deleted it enters the tournament to be a parent of the offspring that will replace it. The algorithm using this policy, along with hypermutation, was able to successfully track the global optimum in both a switching environment, as here, and also in a problem with a continuously moving optimum.

For exercises and recommended reading for this chapter, please visit `www.evolutionarycomputation.org`.

12

Multiobjective Evolutionary Algorithms

In this chapter we describe the application of evolutionary techniques to a particular class of problems, namely multiobjective optimisation. We begin by introducing this class of problems and the particularly important notion of Pareto optimality. We then look at some of the current state-of-the-art multiobjective EAs (MOEAs) for this class of problems and examine the ways in which they make use of concepts of different evolutionary spaces and techniques for promoting and preserving diversity within the population.

12.1 Multiobjective Optimisation Problems

In the majority of our discussions in previous chapters we have made free use of analogies such as adaptive landscapes under the assumption that the goal of the EA in an optimisation problem is to find a single solution that maximises a fitness value that is directly related to a single underlying measure of quality. We also discussed a number of modifications to EAs that are aimed at preserving diversity so that a *set* of solutions is maintained; these represent niches of high fitness, but we have still maintained the conceptual link to an adaptive landscape defined via the assignment of a *single* quality metric (objective) to each of the set of possible solutions.

We now turn our attention to a class of problems that are currently receiving a lot of interest within the optimisation community, and in practical applications. These are the so-called **multiobjective problems** (MOPs), where the quality of a solution is defined by its performance in relation to several, possibly conflicting, objectives. In practice it turns out that a great many applications that have traditionally been tackled by defining a single objective function (quality function) have at their heart a multiobjective problem that has been transformed into a single-objective function in order to make optimisation tractable.

To give a simple illustration (inspired by [334]), imagine that we have moved to a new city and are in the process of looking for a house to buy. There are a number of factors that we will probably wish to take into account, such as: number of rooms, style of architecture, commuting distance to work and method, provision of local amenities, access to pleasant countryside, and of course, price. Many of these factors work against each other (particularly price), and so the final decision will almost inevitably involve a compromise, based on trading off the house's rating on different factors.

The example we have just presented is a particularly subjective one, with some factors that are hard to quantify numerically. It does exhibit a feature that is common to multiobjective problems, namely that it is desirable to present the user with a diverse set of possible solutions, representing a range of different trade-offs between objectives.

The alternative is to assign a numerical quality function to each objective, and then combine these scores into a single fitness score using some (usually fixed) weighting. This approach, often called **scalarisation**, has been used for many years within the operations research and heuristic optimisation communities (see [86, 110] for good reviews), but suffers from a number of drawbacks:

- the use of a weighting function implicitly assumes that we can capture all of the user's preferences, even before we know what range of possible solutions exist.
- for applications where we are repeatedly solving different instances of the same problem, the use of a weighting function assumes that the user's preferences remain static, unless we explicitly seek a new weighting every time.

For these reasons optimisation methods that simultaneously find a *diverse* set of high-quality solutions are attracting increasing interest.

12.2 Dominance and Pareto Optimality

The concept of **dominance** is a simple one: given two solutions, both of which have scores according to some set of objective values (which, without loss of generality, we will assume to be maximised), one solution is said to dominate the other if its score is at least as high for all objectives, and is strictly higher for at least one. We can represent the scores that a solution A gets for n objectives as an n-dimensional vector $\overline{a}$. Using the $\succeq$ symbol to indicate domination, we can define $A \succeq B$ formally as:

$$A \succeq B \iff \forall i \in \{1, \ldots, n\} \ a_i \geq b_i, \text{ and } \exists i \in \{1, \ldots, n\}, \ a_i > b_i.$$

For conflicting objectives, there exists no single solution that dominates all others, and we will call a solution **nondominated** if it is not dominated by any other. All nondominated solutions possess the attribute that their quality cannot be increased with respect to any of the objective functions without detrimentally affecting one of the others. In the presence of constraints, such solutions usually lie on the edge of the feasible regions of the search space. The set of all nondominated solutions is called the **Pareto set** or the Pareto front.

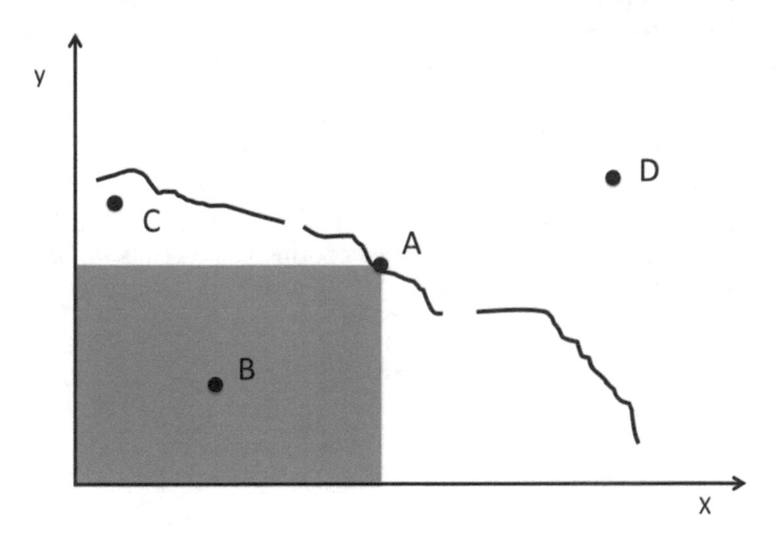

Fig. 12.1. Illustration of the Pareto front. The x- and y-axes represent two conflicting objectives subject to constraints. The quality of solutions is represented by their x and y values (larger is better). Point A dominates B and all other points in the grey area. A and C do not dominate each other. The line represents the Pareto set, of which point A is an example. Solutions above and to the right of the line, such as D, are infeasible

In Figure 12.1 this front is illustrated for two conflicting objectives that are both to be maximised. This figure also illustrates some of the features, such as nonconvexity and discontinuities, frequently observed in real applications that can cause particular problems for traditional optimisation techniques using often sophisticated variants of scalarisation to identify the Pareto set. EAs have a proven ability to identify high-quality solutions in high-dimensional search spaces containing difficult features such as discontinuities and multiple constraints. When coupled with their population-based nature and their ability for finding and preserving diverse sets of good solutions, it is not surprising that EA-based methods are currently the state of the art in many multiobjective optimisation problems.

12.3 EA Approaches to Multiobjective Optimisation

There have been many approaches to multiobjective optimisation using EAs, beginning with Schaffer's vector-evaluated genetic algorithm (VEGA) in 1984 [364]. In this algorithm the population was randomly divided into subpopulations that were then each assigned a fitness (and subject to selection) according to a different objective function, but parent selection and recombination were performed globally. This modification was shown to be enough to preserve an approximation to the Pareto front for a few generations, but not indefinitely.

Subsequent to this, Goldberg suggested the use of fitness based on dominance rather than on absolute objective scores [189], coupled with niching and/or speciation methods to preserve diversity, and this breakthrough triggered a dramatic increase in research activity in this area. We briefly describe some of the best-known algorithms below, noting that the choice of representation, and hence variation operators, are entirely problem dependent, and so we concentrate on the way that fitness assignment and selection are performed.

12.3.1 Nonelitist Approaches

Amongst the first algorithms to explicitly exert selection pressure towards the discovery of nondominated solutions are discussed below:

Fonseca and Fleming's multiobjective genetic algorithm (MOGA) [175] assigns a raw fitness to each solution equal to the number of members of the current population that it dominates, plus one. It uses fitness sharing amongst solutions of the same rank, coupled with fitness-proportionate selection to help promote diversity.

Srinivas and Deb's nondominated sorting genetic algorithm (NSGA) [417] works in a similar way, but assigns fitness based on dividing the population into a number of fronts of equal domination. To achieve this, the algorithm iteratively seeks all the nondominated points in the population that have not been labelled as belonging to a previous front. It then labels the new set as belonging to the current front, and increments the front count, repeating until all solutions have been labelled. Each point in a given front gets as its raw fitness the count of all solutions in inferior fronts. Again fitness sharing is implemented to promote diversity, but this time it is calculated considering only members from that individual's front.

Horn et al.'s niched Pareto genetic algorithm (NPGA) [223] differs in that it uses a modified version of tournament selection rather than fitness proportionate with sharing. The tournament operator works by comparing two solutions first on the basis of whether they dominate each other, and then second on the number of similar solutions already in the new population.

Although all three of these algorithms show good performance on a number of test problems, they share two common features. The first of these is that the performance they achieve is heavily dependent on a suitable choice of parameters in the sharing/niching procedures. The second is that they can potentially lose good solutions.

12.3.2 Elitist Approaches

During the 1990s much work was done elsewhere in the EA research community, developing methods for reducing dependence on parameter settings (Chaps. 7 and 8). Theoretical breakthroughs were achieved showing that single-objective EAs converge to the global optimum on some problems, providing that an elitist strategy (Sect. 5.3.2) is used. In the light of this research Deb and coworkers proposed the revised NSGA-II [112], which still uses the idea of non-dominated fronts, but incorporates the following changes:

- A crowding distance metric is defined for each point as the average side length of the cuboid defined by its nearest neighbours in the same front. The larger this value, the fewer solutions reside in the vicinity of the point.
- A $(\mu + \lambda)$ survivor selection strategy is used (with $\mu = \lambda$). The two populations are merged and fronts assigned. The new population is obtained by accepting individuals from progressively inferior fronts until it is full. If not all of the individuals in the last front considered can be accepted, they are chosen on the basis of their crowding distance.
- Parent selection uses a modified tournament operator that considers first dominance rank then crowding distance.

As can be seen, this achieves elitism (via the plus strategy) and an explicit diversity maintenance scheme, as well as reduced dependence on parameters.

Two other prominent algorithms, the strength Pareto evolutionary algorithm (SPEA-2) [475] and the Pareto archived evolutionary strategy (PAES) [251], both achieve the elitist effect in a slightly different way by using an archive containing a fixed number of nondominated points discovered during the search process. Both maintain a fixed sized archive, and consider the number of archived points close to a new solution, as well as dominance information, when updating the archive.

12.3.3 Diversity Maintenance in MOEAs

To finish our discussion on MOEAs it is appropriate to consider how sets of diverse solutions can be maintained during evolution. It should be clear from the descriptions of the MOEAs above that all of them use explicit methods to enforce preservation of diversity, rather than relying simply on implicit measures such as parallelism (in one form or another) or artificial speciation.

In single-objective optimisation, explicit diversity maintenance methods are often combined with implicit speciation methods to permit the search for optimal solutions within the preserved niches. The outcome of this is a *few* highly fit diverse solutions, often with multiple copies of each (Fig. 5.4). In contrast to this, the aim of MOEAs is to attempt to distribute the population *evenly* along the current approximation to the Pareto front. This partially explains why speciation techniques have not been used in conjunction with the explicit measures. Finally, it is worth noting that the more modern algorithms discussed have abandoned fitness sharing in favour of direct measures of the distance to the nearest nondominating solution, more akin to crowding.

12.3.4 Decomposition-Based Approaches

An unavoidable problem of trying to evenly represent an approximation of the Pareto front is that this approach does not scale well as the dimensionality of the solution space increases above 5–10 objectives. One recent method that has gathered a lot of attention is the decomposition approach taken by Zhang's MOEA-D algorithm [474] which shares features from the single-objective weighted sum approach and the population-based approaches. Rather than use a single weighted combination, MOEA-D starts by evenly distributing a set of N weight vectors in the objective space, and for each building a list of its T closest neighbours (measured by Euclidean distances between them). It then creates and evolves a population of N individuals, each associated with one of the weight vectors, and uses it to calculate a single fitness value. What differentiates this from simply being N parallel independent searches is the use of the neighbourhood sets to structure the population, with selection and recombination only happening within demes, as in cellular EAs (Sect. 5.5.7). Periodically the location of the weight vectors and the neighbour sets are recalculated so that the focus of the N parallel searches reflects what has been discovered about the solution space. MOEA-D, and variants on the original algorithm, have been shown to perform equivalently to dominance-based methods in low dimensions, and to scale far better to many-objective problems with 5 or more conflicting dimensions.

12.4 Example Application: Distributed Coevolution of Job Shop Schedules

An interesting application, which makes use of many of the ideas in this chapter, and also some in Chap. 15, can be seen in Husbands' distributed coevolutionary approach to multiobjective problems [225]. In this approach he uses a coevolutionary model to tackle a complex multiobjective, multi-constraint problem, namely a generalised version of job shop scheduling. Here a number of items need to be manufactured, each requiring a number of operations on different machines. Each item may need a different number of operations, and in general the order of the operations may be varied, so that the problem of finding an optimal production plan for *one* item is itself NP-hard. The usual approach to the multiple task problem is to optimise each plan individually, and then use a heuristic scheduler to interleave the plans so as to obtain an overall schedule. However, this approach is inherently flawed because it optimises the plans in isolation rather than taking into consideration the availability of machines, etc.

Husbands' approach is different: he uses a separate population to evolve plans for each item and optimises these concurrently. In this sense we have a MOP, although the desired final output is a single set of plans (one for each item) rather than a set of diverse schedules. A candidate plan for one item gets evaluated in the context of a member from each other population, i.e., the fitness value (related to time and machining costs) is for a complete production schedule. An additional population is used to evolve 'arbitrators', which resolve conflicts during the production of the complete schedule.

Early experiments experienced problems with premature loss of diversity. These problems are treated by the use of an implicit approach to diversity preservation, namely the use of a diffusion model EA. Furthermore, by colocating one individual from each population in each grid location, the problem of partner selection (Sect. 15.2.1) is neatly solved: a complete solution for evaluation corresponds to a grid cell.

We will not give details of Husbands' representation and variation operators here, as these are highly problem specific. Rather we will focus on the details of his algorithm that were aimed at aiding the search for high-class solutions. The first of these is, of course, the use of a coevolutionary approach. If a single population were used, with a solution representing the plans for all items, there would be a greater likelihood of genetic hitchhiking (see Sect. 16.1 for a description), whereby a good plan for one item in the initial population would take over, even if the plans for the other items were poor. By contrast, the decomposition into different subpopulations means that the good plan can at worst take over one population.

The second feature that aids the search over diverse local optima is the use of a diffusion model approach. The implementation uses a 15-by-15 square toroidal grid, thus a population size of 225. Plans for 5 items, each needing between 20 and 60 operations, were evolved, so in total there were 6 populations, and each cell contained a plan for each of the 5 items plus an arbitrator. A generational approach is used: within each generation each cell's populations are 'bred', with a random permutation to decide the order in which cells are considered.

The breeding process within each cell is iterated for each population and consists of the following steps:

1. Generate a set of points to act as neighbours by iteratively generating random lateral and vertical offsets from the current position. A binomial approximation to a Gaussian distribution is used, which falls off sharply for distances more than 2 and is truncated to distance 4.
2. Rank the cells in this neighbourhood according to cost, and select one using linear ranking with $s = 2$.

3. Take the member of the current population from the selected cell and the member in the current cell, and generate an offspring via recombination and mutation.
4. Choose a cell from the neighbourhood using inverse linear ranking.
5. Replace the member of the current population in that cell with the newly created offspring.
6. Re-evaluate all the individuals in that cell using the newly created offspring.

The results presented from this technique showed that the system managed to evolve low-cost plans for each item, together with a low total schedule time. Notably, even after several thousand iterations, the system had still preserved a number of diverse solutions.

For exercises and recommended reading for this chapter, please visit `www.evolutionarycomputation.org`.

13

Constraint Handling

In this chapter we return to an issue first introduced in Sect. 1.3, namely that some problems have constraints associated with them. This means that not all possible combinations of variable values represent valid solutions to the problem at hand, and we examine how this impacts on the design of an evolutionary algorithm. This issue has great practical relevance because many real-world problems are constrained. It is also theoretically challenging, since many intractable problems (NP-hard, NP-complete, etc.) are constrained. Unfortunately, constraint handling is not straightforward in an EA, because the variation operators (mutation and recombination) are typically 'blind' to constraints. This means that even if the parents satisfy some constraints, there is no guarantee their offspring will. This chapter reviews the most commonly used techniques for constraint handling, identifies a number of common features, and provides some guidance for the algorithm designer.

13.1 Two Main Types of Constraint Handling

Before discussing how constraints may be dealt with, we first briefly recap our classification from Chap. 1. That was based on whether problems contained two features: constraints on the form that solutions were allowed to take; and a quality, or fitness function. If we have:

- neither feature then we do not have a problem;
- a fitness function but no constraints then we have a Free Optimisation Problem (FOP);
- constraints that a candidate solution must meet but no other fitness criteria then we have a Constraint Satisfaction Problem (CSP);
- both a fitness function and constraints then we have a Constrained Optimisation Problem (COP).

Finally, we discussed how, depending on how we frame a problem, we might have different numbers of possible solutions. In Chap. 1 we illustrated this using the eight-queens problem, restricting the search space of a CSP to give a much reduced COP, which is far easier to solve. Here we attempt to broaden this concept. To that end, various techniques for constraint handling are discussed in Sect. 13.2. Before going into details, let us distinguish two conceptually different possibilities.

- In the case of **indirect constraint handling**, constraints are transformed into optimisation objectives. After the transformation, they effectively disappear, and all we need to care about is optimising the resulting objective function. This type of constraint handling is done *before* the EA run.
- In **direct constraint handling**, the problem offered to the EA to solve has constraints (is a COP) that are enforced explicitly *during* the EA run.

These options are not exclusive: for a given constrained problem (CSP or COP) some constraints might be treated directly and some others indirectly.

In fact, even when all constraints are treated indirectly, so that our EA is applied to a FOP, this does not mean that the EA is necessarily ignoring the constraints. In theory one could fully rely on the general optimisation power of EAs and try to solve the given FOP without taking note of how the values of the fitness function f are obtained. It would remain the designer's responsibility (and one of the main design guidelines) to ensure that solutions to the transformed FOP represent solutions to the original CSP or COP. However, it is also possible to take the origin of f into account, i.e., the fact that it is constructed from constraints. For example, one can try to make use of specific constraint-based information within the EA by designing special mutation or crossover operators that explicitly use heuristics to try to ensure that offspring satisfy more constraints than the parents.

13.2 Approaches to Handling Constraints

In the discussion so far, we have not considered the nature of the domains of the variables. In this respect there are two extremes: they are all discrete or all continuous. Continuous CSPs are rather rare, so by default a CSP is discrete [433]. For COPs this is not the case as we have discrete COPs (**combinatorial optimisation** problems) and continuous COPs as well. Much of the evolutionary literature on constraint handling is restricted to one of these cases, but in fact the ways for handling constraints are practically identical – at least at the conceptual level. Therefore the following treatment of constraint handling methods is general, and we note simply that the presence of constraints will divide the space of potential solutions **S** into two or more disjoint regions, the **feasible region** (or regions) **F** containing those candidate solutions that satisfy the given constraints, and **U**, the **infeasible region** containing those that do not.

We distinguish between approaches by considering how they modify one or more facets of the search: the genotype space, the phenotype space **S**, the mapping from genotype to phenotype, or the fitness function. Building on our division above, the commonly accepted set of methods used is:

- Indirect Approaches
  1. Penalty functions modify the fitness function. For feasible solutions in **F** the objective function values are used, but for those in **U** an extra penalty value is applied. Preferably this is designed so that the fitness is reduced in proportion to the number of constraints violated, or to the distance from the feasible region.
- Direct Approaches
  1. Specialised representations, together with initialisation, and reproduction operators reduce the genotype space to ensure all candidate solutions are feasible. The mapping, phenotype space and fitness functions are left unchanged. The use of permutation representations with specialised recombination and mutation as described in Sect. 4.5 is an example of this approach. Although this example is fairly straightforward, in more complex cases it may be hard to reverse-engineer the mapping in order to ensure that all of the valid phenotype space is covered by the new smaller genotype space.
  2. Repair mechanisms modify the original mapping. For feasible solutions in **F** the mapping is left unchanged, but for those in **U** an extra stage is added to turn an infeasible solution into a feasible one, hopefully close to the infeasible one. Note that this assumes that we can in some sense evaluate a solution to see if it violates constraints.
  3. Decoder functions that replace the original mapping from genotype to phenotype so that all solutions (i.e., phenotypes) are guaranteed to be feasible. The genotype space and fitness functions are left unchanged, and standard evolutionary operators may be applied. Unlike repair functions, which work on whole solutions, decoder functions typically take constraints into account as a solution is constructed from partial components.

In the following sections we briefly discuss the above approaches, focusing on the facets that have implications for the applications of EAs in general. We will use as a common example the 0–1 knapsack problem introduced in Sect. 3.4.2. We are given a set of n items, each with an associated value and cost, $v(i), sc(i) : 1 \leq i \leq n$. The usual representation is a binary vector $\overline{x} \in \{0,1\}^n$, and we seek the vector $\overline{x}*$ which maximises the value of the chosen items $\sum_i x(i) \cdot v(i)$ subject to the maximum cost constraint $\sum_i x(i) \cdot c(i) \leq C_{max}$.

We should note that, in practice, it is common to utilise as much domain-specific knowledge as possible, in order to reduce the amount of time spent generating infeasible solutions. As is pointed out in [302], the global optimum of a COP with continuous variables often lies on, or very near to, the boundary between the feasible and infeasible regions, and promising results are reported using algorithms that specifically search along that boundary.

However, we concentrate here on the more general case, since the domain knowledge required to specify such operators may not be present.

For a fuller review of work in this area, the reader is referred to [127, 141, 300, 302]. Furthermore, [90, 298, 301, 361] are especially recommended because they contain descriptions of problem instance generators for binary CSPs [90], continuous COPs [298, 301], or a large collection of continuous COP test landscapes [361], together with detailed experimental results. Reinforcing this book's emphasis on the importance of choosing appropriate representations, [302] reported that for problems in the continuous domain, use of a real-valued rather than binary representation consistently gave better results.

13.2.1 Penalty Functions

Penalty functions modify the original fitness function $f(\overline{x})$ applied to a candidate solution $\overline{x}$ such that $f'(\overline{x}) = f(\overline{x}) + P(d(\overline{x}, F))$, where $d(\overline{x}, F))$ is a distance metric of the infeasible point to the feasible region F (this might be simply a count of the number of constraints violated). The penalty function P is zero for feasible solutions, and it increases with distance from the feasible region (for minimisation problems).

For our knapsack problem in Sect. 3.4.2, one simple approach is to calculate the excess weight $e(\overline{x}) = \sum_i x(i) \cdot c(i) - C_{max}$, and then use the penalty function:

$$P(\overline{x}) = \begin{cases} 0 & \mathit{if} \quad e(\overline{x}) \leq 0 \\ w \cdot e(\overline{x}) & \mathit{if} \quad e(\overline{x}) > 0. \end{cases}$$

where the fixed weight w is large enough that feasible solutions are preferred.

It is important to note that this approach assumes that it is possible to evaluate an infeasible point; although in this example it is, for many others this is not the case. This discussion is also confined to *exterior* penalty functions, where the penalty is only applied to infeasible solutions, rather than *interior* penalty functions, which apply penalties to all solutions based on distance from the constraint boundary in order to encourage exploration of this region.

The conceptual simplicity of penalty function methods means that they are widely used, and they are especially suited to problems with disjoint feasible regions, or where the global optimum lies on (or near) the constraint boundary. However, their successful use depends on a balance between exploration of the infeasible region and not wasting time, which places a lot of emphasis on the form of the penalty function and the distance metric.

If the penalty function is too severe, then infeasible points near the constraint boundary will be discarded, which may delay, or even prevent, exploration of this region. Equally, if the penalty function is not sufficient in magnitude, then solutions in infeasible regions may dominate those in feasible regions, leading to the algorithm spending too much time in the infeasible regions and possibly stagnating there. In general, for a system with m constraints, the form of the penalty function is a weighted sum

$$P(d(\overline{x}, F)) = \sum_{i=1}^{m} w_i \cdot d_i^{\kappa}(\overline{x})$$

where κ is a user-defined constant, often taking the value 1 or 2, and as above the distance metrics $d_i(\overline{x})$ from the point $\overline{x}$ to the boundary for constraint i may be a simple binary value according to whether the constraint is satisfied, or a metric based on cost of repair.

Many different approaches have been proposed, and a good review is given in [379], where penalty functions are classified as *constant, static, dynamic,* or *adaptive.* This classification closely matches the options discussed in the example given in Sect. 8.2.2.

Static Penalty Functions

Three methods have commonly been used with static penalty functions, namely *extinctive* penalties (where all of the w_i are set so high as to prevent the use of infeasible solutions), binary penalties (where the value d_i is 1 if the constraint is violated, and zero otherwise), and distance-based penalties.

It has been reported that, of these three, the latter give the best results [189], and the literature contains many examples of this approach. This approach relies on the ability to specify a distance metric that accurately reflects the difficulty of repairing the solution, which is obviously problem dependent, and may also vary from constraint to constraint. The usual approach is to take the square of the Euclidean distance (i.e., set $\kappa = 2$) .

However, the main problem in using static penalty functions remains the setting of the values of w_i. In some situations it may be possible to find these by experimentation, using repeated runs and incorporating domain-specific knowledge, but this is a time-consuming process that is not always possible.

Dynamic Penalty Functions

An alternative approach to setting fixed values of w_i by hand is to use dynamic values, which vary as a function of time. A typical approach is that of [237], in which the static values w_i were replaced with a simple function of the form $s_i(t) = (w_i t)^{\alpha}$, where it was found that for best performance $\alpha \in \{1, 2\}$. Although this approach is possibly less brittle as a result of not using fixed (possibly inappropriate) values for w_i, the user must still decide on the initial values.

An alternative, which can be seen as the logical extension of this approach, is the behavioural memory algorithm of [369]. Here a population is evolved in a number of stages – the same number as there are constraints. In each stage i, the fitness function used to evaluate the population is a combination of the distance function for constraint i with a death penalty for all solutions violating constraints $j < i$. In the final stage all constraints are active, and the objective function is used as the fitness function. It should be noted that different results may be obtained, depending on the order in which the constraints are dealt with.

Adaptive Penalty Functions

Adaptive penalty functions represent an attempt to remove the danger of poor performance resulting from an inappropriate choice of values for the penalty weights w_i. An early approach described in [45, 205] was discussed in Sect. 8.2.2. A second approach is that of [380, 426], in which adaptive scaling (based on population statistics of the best feasible and infeasible raw fitnesses yet discovered) is coupled with the distance metrics for each constraint based on the notion of "near feasible thresholds". These latter are scaling factors for each distance metric, which can vary with time.

The Stepwise Adaptation of Weights (SAW) algorithm of [149, 150, 151] can be seen as a population-level adaptation of the search space. In this method the weights w_i are adapted according to a simple heuristic: if the best individual in the current population violates constraint i, then this constraint must be hard and its weight should be increased. In contrast to the adaptive mechanisms above, the updating function is much simpler. In this case a fixed penalty increment Δw is added to the penalty values for each of the constraints violated in the best individual of the generation at which the updating takes place. This algorithm was able to adapt weight values that were independent of the EA operators and the initial weight values, suggesting that this is a robust technique.

13.2.2 Repair Functions

The use of repair algorithms for solving COPs with EAs can be seen as a special case of adding local search to the EA. In this case the aim of the local search is to reduce (or remove) the constraint violation, rather than to simply improve the value of the fitness function, as is usually the case.

The use of local search has been intensively researched, with attention focusing on the benefits of so-called Baldwinian versus Lamarckian learning (Sect. 10.2.1). In either case, the repair algorithm works by taking an infeasible point and generating a feasible solution based on it. In the Baldwinian case, the fitness of the repaired solution is allocated to the infeasible point, which is kept, whereas with Lamarckian learning, the infeasible solution is overwritten with the new feasible point. Although the Baldwin vs. Lamarck debate has not been settled within unconstrained learning, many COP algorithms reach a compromise by introducing some stochasticity, for example Michalewicz's GENOCOP algorithm uses the repaired solution around 15% of the time [299].

For our knapsack example, a simple repair method is to change some of the gene values in $\overline{x}$ from 1 to 0. Although this sounds simple, this example raises some interesting questions. One of these is the replacement question just discussed; the second is whether the genes should be selected for altering in a predetermined order, or at random. In [295] it was reported that using a greedy deterministic repair algorithm gave the best results, and certainly the use of a nondeterministic repair algorithm will add noise to the evaluation of every individual, since the same potential solution may yield different fitnesses on separate evaluations. However, it has been found by some authors [398] that the addition of noise can assist the GA in avoiding premature convergence. In practice it is likely that the best method is not only dependent on the problem instance, but on the size of the population and the selection pressure.

Although the knapsack example is fairly simple, in general defining a repair function may be as complex as solving the problem itself. One algorithm that eases this problem (and incidentally uses stochastic repair), is Michalewicz's GENOCOP III algorithm for optimisation in continuous domains [299]. This works by maintaining two populations, one P_s of so-called search points and one P_r of 'reference points', with all of the latter being feasible. Points in P_r and feasible points from P_s are evaluated directly. When an infeasible point is generated in P_s it is repaired by picking a point in P_r and drawing a line segment from it to the infeasible point. This is then sampled until a repaired feasible point is found. If the new point is superior to that used from P_r, the new point replaces it. With a small probability (which represents the balance between Lamarckian and Baldwinian search) the new point replaces the infeasible point in P_s. It is worth noting that although two different methods are available for selecting the reference point used in the repair, both are stochastic, so the evaluation is necessarily noisy.

13.2.3 Restricting Search to the Feasible Region

In many COP applications it may be possible to construct a representation and operators so that the search is confined to the feasible region of the search space. In constructing such an algorithm, care must be taken in order to ensure that all of the feasible region is capable of being represented. It is equally desirable that any feasible solution can be reached from any other by (possibly repeated) applications of the mutation operator. The classic example of this is permutation problems. In Sect. 3.4.1 we showed an illustration for the eight-queens problem, and in Sects. 4.5.1 and 4.5.2 we described a number of variation operators that deliver feasible offspring from feasible parents.

For our knapsack problem, we could imagine the following operators. A randomised initialisation operator might construct solutions by starting with an empty set $x(i) = 0, \forall i$ and randomly picking elements i to flip the gene value from to 1 until adding the next value chosen would violate the cost constraint. This would give an initial population where the excess cost $e(\overline{x})$ was negative for each member. For recombination, we could apply a slightly modified one-point crossover. For any given pair of parents, first we generate a random permutation of the values $\{1,, \ldots, n{-}1\}$ in which to consider the potential crossover points. In that order we consider the pairs of offspring created, accepting the first pair that is feasible. For mutation we apply bitwise mutation, accepting any move that changes a gene from 1 to 0, but only those from 0 to 1 that do not created excess cost. Again we might choose to do this in a random order to remove bias towards selecting items at the start of our representation.

It should be noted that this approach to solving COP, although attractive, is not suitable for all types of constraints. In many cases it is difficult to find an existing or design a new operator that guarantees that the offspring are feasible. Although one possible option is simply to discard any infeasible points and reapply the operator until a feasible solution is generated, the process of checking that a solution is feasible may be so time consuming as to render this approach unsuitable. However, there remains a large class of problems where this approach is valid and with suitable choice of operators can be very successfully applied.

13.2.4 Decoder Functions

Decoder functions are a class of mappings from the genotype space S' to the feasible regions F of the solution space S that have the following properties:

- Every $z \in S'$ must map to a *single* solution $s \in F$.
- Every solution $s \in F$ must have at least one representation $s' \in S'$.
- Every $s \in F$ must have the same number of representations in S' (this need not be 1).

Such decoder functions provide a relatively simple way of using EAs for this type of problem, but they are not without drawbacks. These are centred around the fact that decoder functions generally introduce a lot of redundancy into the original genotype space. This arises when the new mapping is many-to-one, meaning that a number of potentially radically different genotypes may be mapped onto the same phenotype, and only a subset of the phenotype space can be reached.

Considering the knapsack example, a simple approach would leave the genotype, initialisation and variation operators unchanged. When constructing a solution, the decoder function could start at the left hand end of the string and interpret a 1 as *take this item if possible* ... If the cost limit is reached after considering, say, j of the n genes, then it is irrelevant what values the rest take, and so 2^{n-j} strings all map onto the same solution.

In a few cases it may be possible to devise a decoder function that permits the use of relatively standard representation and operators while preserving a one-to-one mapping between genotype and phenotype. One such example is the decoder for the TSP problem proposed by Grefenstette, which is well described by Michalewicz in [297]. In this case a simple integer representation was used with each gene $a_i \in \{1, \ldots, n+1-i\}$. This representation permits the use of common crossover operators and a bitwise mutation operator that randomly resets a gene value to one of its permitted allele values. The outcome of both of these operators is guaranteed to be valid. The decoder function works by considering an ordered list of cities, $ABCDE$, and using the genotype to index into this.

For example, with a genotype $\langle 4,2,3,1,1 \rangle$ the first city in the constructed tour is the fourth item in the list, i.e., D. This city is then removed from the list and the second gene is considered, which in this case points to B. This process is continued until a complete tour is constructed: $\langle 4,2,3,1,1 \rangle \rightarrow DBEAC$.

Although the one-to-one mapping means that there is no redundancy in the genotype space, and it permits the use of straightforward crossover and mutation operators, the complexity of the mapping function means that a small mutation can have a large effect, e.g., $\langle 3,2,3,1,1 \rangle \rightarrow CBDAE$. Equally, it can be easily shown that recombination operators no longer respect and propagate all features common to both solutions. Thus if the two solutions $\langle 1,1,1,1,1 \rangle \rightarrow ABCDE$ and $\langle 5,1,2,3,1 \rangle \rightarrow EACDB$, which share the common feature that C occurs in the third position and D in the fourth undergo 1-point crossover between the third and fourth loci, the solution $\langle 5,1,2,1,1 \rangle \rightarrow EACBD$ is obtained, which does not possess this feature. If the crossover occurs in other positions, the edge CD may be preserved, but in a different position in the cycle.

In both of the examples given, the complexity of the genotype–phenotype mapping makes it very difficult to ensure locality and makes the fitness landscape associated with the search space highly complex, since the potential effects in fitness of changes at the left-hand end of the string are much bigger than those at the right-hand end [196]. Equally, it can become very difficult to specify exactly the common features the recombination operators are supposed to be preserving.

13.3 Example Application: Graph Three-Colouring

We illustrate the approaches outlined above via the description of two different ways of solving a well-known CSP problem, graph three-colouring. This is an abstract version of colouring a political map so that no two adjacent areas (counties, states, countries) have the same colour. We are given a graph $G = \{v, e\}$ with $n = |v|$ vertices and $m = |e|$ edges connecting some pairs of the vertices. The task is to find, if possible, an assignment of one of three colours to each vertex so that there are no edges in the graph connecting same-coloured vertices.

Indirect Approach

We begin by illustrating an indirect approach, transforming the problem from a CSP to a FOP by means of penalty functions. The most straightforward representation is using ternary strings of length $n = |v|$, where each variable stands for one node, and the integers 1, 2, and 3 denote the three colours.

Using this standard GA representation has the advantage that all standard variation operators are immediately applicable. We now define two objective functions (penalty functions) that measure the amount of 'incorrectness' of a chromosome. The first function is based on the number of 'incorrect edges' that connect two nodes with the same colour, while the second relies on counting the 'incorrect nodes' that have a neighbour with the same colour. For a formal description let us denote the constraints belonging to the edges as c_i ($i = \{1, \ldots, m\}$), and let C^i be the set of constraints involving variable v_i (edges connecting to node i). Then the penalties belonging to the two options described above can be expressed as follows:

$$f_1(\bar{s}) = \sum_{i=1}^{m} w_i \times \chi(\bar{s}, c_i),$$

where $\chi(\bar{s}, c_i) = \begin{cases} 1 & \text{if } \bar{s} \text{ violates } c_i, \\ 0 & \text{otherwise.} \end{cases}$

Respectively,

$$f_2(\bar{s}) = \sum_{i=1}^{n} w_i \times \chi(\bar{s}, C^i),$$

where $\chi(\bar{s}, C^i) = \begin{cases} 1 & \text{if } \bar{s} \text{ violates at least one } c \in C^i, \\ 0 & \text{otherwise.} \end{cases}$

Note that both functions are correct transformations of the constraints in the sense that for each $\bar{s} \in S$ we have that $\phi(\bar{s}) = true$ if and only if $f_i(\bar{s}) = 0$ ($i = 1, 2$). The motivation to use weighted sums in this example, and in general, is that they provide the possibility of emphazising certain constraints (variables) by giving them a higher weight. This can be beneficial if some constraints are more important or known to be harder to satisfy. Assigning them a higher weight gives a higher reward to a chromosome, hence the EA naturally focuses on these. Setting the weights can be done manually by the user, but can also be done by the EA itself on-the-fly as in the stepwise adaptation of weights (SAW) mechanism [151].

Now the EA for the graph three-colouring problem can be composed from standard components. For instance, we can apply a steady-state GA with population size 100, binary tournament selection and worst fitness deletion, using random resetting mutation with $p_m = 1/n$ and uniform crossover with $p_c = 0.8$. Notice that this EA really ignores constraints; it only tries to minimise the given objective function (penalty function).

Direct Approach

For this problem, two of the direct approaches would be extremely difficult, if not impossible, to implement. Specifying either an initialisation operator, or a repair function, to create valid solutions would effectively mean solving the problem, and since it is thought to be *NP*-complete it is unlikely that there is a polynomial time algorithm that could accomplish either of these.

However, we now present another EA for this problem, illustrating how constraints can be handled by a decoder. The main idea is to use permutations of the nodes as chromosomes. The phenotype (colouring) belonging to a genotype (permutation) is determined by a procedure that assigns colours to nodes in the order they occur in the given permutation, trying the colours in increasing order (1,2,3), and leaving the node uncoloured if all three colours would lead to a constraint violation. Formally, we shift from the search space of all colourings $S = \{1,2,3\}^n$ to the space of all n-long permutations $S' = \{\bar{s} \in \{1,\ldots,n\}^n \mid s_i \neq s_j \;\; i,j = 1,\ldots,n\}$, and the colouring procedure (the decoder) is the mapping from S' to S. At first glance this might not seem like a good idea as we still have constraints in the transformed problem – those that define the property of being a permutation in the definition of S' above. However, we know from Sect. 4.5 that working in a permutation space is easy, as there are many suitable variation operators keeping the search in this space. In other words, we have various operators preserving the constraints defining this space.

An appropriate objective function for this representation can simply be defined as the number (weighted sum) of nodes that remain uncoloured after decoding. This function also has the property that an optimal value (0) implies that all constraints are satisfied, i.e., all nodes are coloured correctly. The rest of the EA can again use off-the-shelf components: a steady-state GA with population size 100, binary tournament selection and worst fitness deletion, using swap mutation with $p_m = 1/n$ and order crossover with $p_c = 0.8$.

Looking at this solution at a conceptual level we can note that there are two constraint-handling issues. Primary constraint-handling concerns handling the constraints of the original problem, the graph three-colouring CSP. This is done by the mapping approach via a decoder. However, the transformed search space S' in which the EA has to work is not free, rather it is restricted by the constraints defining permutations. This constitutes the secondary constraint handling issue that is solved by a (direct) preserving approach using appropriate variation operators.

For exercises and recommended reading for this chapter, please visit
`www.evolutionarycomputation.org`.

14

Interactive Evolutionary Algorithms

This chapter discusses the topic of **interactive evolution**, where the measure of a solution's fitness is provided by a human's subjective judgement, rather than by some predefined model of a problem. Of course, the world around us is full of examples of human intervention in biological evolution, in the form of pets, garden flowers, food crops and farm animals. Applications of **Interactive Evolutionary Algorithms** (IEAs) range from capturing aesthetics in art and design, to the personalisation of artefacts such as medical devices. When including humans 'in the loop' we must consider their peculiar characteristics. On one hand, they can provide insight and guidance beyond simply selecting parents for breeding. On the other, they can be inconsistent, and are prone to fatigue and loss of attention. These factors make it inappropriate to use the 'traditional' model of an EA generating possibly thousands of candidate solutions. This chapter describes and explains some of the major algorithmic changes that have been proposed to cope with these issues.

14.1 Characteristics of Interactive Evolution

The defining feature of IEAs is that the user effectively becomes part of the system, acting as a guiding oracle to control the evolutionary process. As a starting point, consider agricultural breeding, where human interference changes the reproductive process. Based on functional (faster horses, higher yields) or aesthetic (nicer cats, brighter flowers) judgements, a supervisor selects the individuals that are allowed to reproduce. Over time, new types of individuals emerge that meet the human expectations better than their ancestors. From this familiar process we can now start to distinguish some particular features that impact on the design of IEAs.

14.1.1 The Effect of Time

Many crop plants have annual life cycles, and farm animals may take years to reach maturity. Thus in plant breeding and animal husbandry we are used to the idea of many humans playing successive roles, in a process that may take centuries. Although both of these are nonstationary problems, since fashions in flowers, pets, and food change, the relative infrequency of intervention, and the importance of making the right choice, makes human input fairly reliable.

However, when we are considering simulated evolution, then many decisions may be required fairly rapidly from a single user. Also, each individual decision may seem less important, since we are used to the idea that we can re-run computer programs. In these cases human fatigue, and its effect on the consistency of evaluation, becomes a major factor, even if the person is attempting to apply a well-understood standard. People have a natural limited attention span, and, like fatigue, a loss of engagement with the process has been shown to lead to increasingly erratic decisions as the user performs more and more evaluations [76]. Thus, from the perspective of evolutionary time, there is a need to avoid lengthy evolution over hundreds or thousands of generations, and instead focus on making rapid gains to fit in with human needs.

Even taking this into account, from the perspective of wall-clock time IEAs can be slow to run, as it usually takes much longer for a person to make a decision than it does to calculate a mathematical fitness function. The net effect of time constraints and human cognitive limitations is that typically only a small fraction of the search space is considered. This is the major challenge facing IEA designers, and it is common for successful IEA applications to employ one or more different approaches to alleviate this issue. A sense of progress can be achieved by evaluating fewer solutions per generation — either via small populations, or by only evaluating some of a large population. Frustration can be reduced by not losing good solutions. More generally, engagement can be increased, and the onset of fatigue delayed, by allowing the user more direct input into the process.

14.1.2 The Effect of Context: What Has Gone Before

Closely related to the issue of time, or the length of the search process, is that of context. By this we mean that human expectations, and ideas about what is a good solution, change in response to what evolution produces. If (as we hope) people are pleasantly surprised by what they see, then their expectations rise. This can mean that a solution might be judged average early in a run, but only sub-standard after more solutions have been presented. Alternatively, after a few generations of viewing similar solutions, a user may decide that they are in a 'blind alley'. In that case they might wish to return to what were previously thought to be only average solutions to explore their potential. Both of these imply a need to generate a diverse range of solutions — either by increasing the rate of mutation or by a restart mechanism, or by maintaining some kind of archive.

14.1.3 Advantages of IEAs

Despite, or perhaps because of, the issues identified above, it is worth reiterating the potential advantages of incorporating humans directly into the evolutionary cycle. These were summarised by Bentley and Corne in [50]:

- Handling situations with no clear fitness function. If the reasons for preferring certain solutions cannot be formalised, no fitness function can be specified and implemented within the EA code. Subjective user selection circumvents this problem. It is also helpful when the objectives and preferences are changeable, as it avoids having to rewrite the fitness function.
- Improved search ability. If evolution gets stuck, the user can redirect search by changing his or her guiding principle.
- Increased exploration and diversity. The longer the user 'plays' with the system, the more and more diverse solutions will be encountered.

14.2 Algorithmic Approaches to the Challenges of IEAs

Having outlined the issues confronting IEA designers, we now discuss some of the ways in which they have been addressed.

14.2.1 Interactive Selection and Population Size

In general, the user can influence selection in various ways. The influence can be very direct, for example, actually choosing the individuals that are allowed to reproduce. Alternatively, it can be indirect — by defining the fitness values or perhaps only sorting the population, and then using one of the selection mechanisms described in Sect. 5.2. In all cases (even in the indirect one) the user's influence is named **subjective selection**, and in an evolutionary art context the term **aesthetic selection** is often used.

Population size can be an important issue for a number of reasons. For visual tasks, computer screens have a fixed size and humans need a certain minimum image resolution, which limits how many solutions can be viewed at once. If the artefacts being evolved are not visual (or if they are, but the population cannot be viewed simultaneously), then the user has to rely heavily on memory to rank or choose individuals. Psychology provides the rule of thumb that people only hold around seven things in memory. Another aspect is that if a screenful of solutions are being ranked, the number of pairwise comparisons and decisions to be made grows with the square of the number onscreen. For all these reasons, interactive EAs frequently work with relatively small populations. Also, small populations can help keep the user engaged by providing them with a sense of progress.

Since the mid-2000s a number of authors have investigated ways of dealing with problems that exhibit a mixture of quantitative and qualitative aspects.

The usual approach is to use a multi-objective EA as described in Chap. 12, and the increased solution space usually requires bigger population sizes. However, not all of these need be evaluated by the user, since some can be ruled out on the basis of inferior performance on quantitative metrics.

14.2.2 Interaction in the Variation Process

However controversially, advances in biology have meant that agricultural breeding can now act not just on the selection process, but also on variation operators — for example, inserting genes to produce genetically modified crops. Similarly, Interactive Artificial Evolution can permit direct user intervention in the variation process. In some cases this is implicit - for example allowing the user to periodically adjust the choice and parameterisation of variation operators. Caleb-Solly and Smith [76] used the score given to an individual to control mutation — so that 'worse' individuals were more likely to have values changed, or to have values changed by a larger amount. As well as noting performance benefits, they also reported an effect which is typical of human interactions with Artificial Intelligence. Specifically, users' behaviour changed over time, as they got used to the system and developed a perception of how it would respond to their input. This manifested as users alternating between *exploiting* promising regions (awarding high scores to cause small changes), and *exploring* new regions (awarding low scores as a kind of reset button).

More explicit forms of control may also be used. For instance, interactive evolutionary timetabling might allow planners to inspect promising solutions, interchange events by hand and place the modified solutions back to the population. This is an explicit Lamarckian influence on variation.

Both types of algorithmic adaptation have in common the desire to maximise the value of each interaction with a user. Methods for directly affecting variation are usually designed with the goal of increasing the rate of adaptation towards good solutions by removing some of the black-box elements of search. However, in practice a nice synergy has been observed – in fact the ability to guide search actually increases user engagement, and so delays the onset of fatigue [335]. Again, this mimics results elsewhere in interactive AI.

14.2.3 Methods for Reducing the Frequency of User Interactions

The third major algorithmic adaptation attempts to reduce the number of solutions a user is asked to evaluate, while simultaneously maintaining large populations and/or using many generations. This is done by use of a **surrogate fitness function** which attempts to approximate the decisions a human would make. In fact this approach is also used to reduce the wall-clock time of evolutionary algorithms working with heavily time-intensive fitness functions.

Typically the surrogate fitness models are adaptive and attempt to learn to reflect the users' decisions. Normally the model is used to provide fitness for all solutions, and only occasionally is the user asked to evaluate a solution. This input can then be used as feedback to update the parameters of the model used. Hopefully, over time, the fitness values predicted by the surrogate model match the users' input better and better. Models used range from the simple (offspring receive the mean fitness of their parents) to the use of advanced machine learning algorithms such as Support Vector Machines and Neural Networks. The crucial decisions from an algorithmic perspective are: how complex a model is necessary; how often should the real fitness function be invoked (in this case a human evaluation); how should solutions be chosen to be properly evaluated; and how should the surrogate model be updated to reflect the new information obtained. Full discussion is beyond the scope of this book, but the interested reader will find a good review of this type of approach in [235] and a discussion of recent developments and the current research issues in [236].

14.3 Interactive Evolution as Design vs. Optimisation

Interactive evolution is often related to evolutionary art and design. It can even be argued that *evolution is design*, rather than optimisation. From this conceptual perspective the canonical task of (natural or computer) evolution is to design good solutions for specific challenges. Many exponents have arrived at a view that distinguishes parameter optimisation and exploration [48, 49], the main underlying difference being the representation of a solution.

Many problems can be solved by defining a parameterised model of possible solutions and seeking the parameter values that encode an optimal solution. This encoding is 'knowledge-rich' in the sense that the appropriate parameters must be chosen intelligently – if there is no parameter for a feature that influences solution quality that feature can never be modified, different values cannot be compared, and possibly good solutions will be overlooked. *Design optimisation* typically uses this type of representation. Propagating changes across generations, an EA acts as an optimiser in the parameter space.

An alternative to parameterised representations is component-based representation. Here a set of low-level components is defined, and solutions are constructed from these components. This is a 'knowledge-lean' representation with possibly no, or only weak, assumptions of relationships between components and the ways they can be assembled. IEAs for evolutionary design commonly use this sort of representation. Also known as **Generative and Developmental Systems**, these quite naturally give rise to exploration, aiming at identifying novel and good solutions, where novelty can be more important than optimality (which might not even be definable). In this paradigm, evolution works as a *design discovery* engine, discovering new designs and helping to identifying new design principles by analysing the evolved designs [303]. The basic template for interactive evolutionary design systems consists of five components [50]:

1. A phenotype definition specifying the application-specific kind of objects we are to evolve.
2. A genotype representation, where the genes represent (directly or indirectly) a variable number of components that make up a phenotype.
3. A decoder, often called growth function or **embryogeny**, defining the mapping process from genotypes to phenotypes.
4. A solution evaluation facility allowing the user to perform selection within the evolutionary cycle in an interactive way.
5. An evolutionary algorithm to carry out the search.

This scheme can be used to evolve objects with an amazing variety, including Dawkins' pioneering Biomorphs [101], coffee tables [47], images imitating works of the artist M.C. Escher [138], scenes of medieval towns [405], music [307], and art such as the 'Mondriaan Evolver' [437] or collaborative online 'art breeder' systems.[1] The basics of such evolutionary art systems are the same as those of evolutionary design in general: some evolvable genotype is specified that encodes an interpretable phenotype. These might be visual (two-dimensional images, animations, and three-dimensional objects) or audible (sounds, ringtones, music). The main feature that distinguishes the application of IEAs to art from other forms of design lies in the intention: the evolved objects are simply to please the user, and need not serve any practical purpose [355].

14.4 Example Application: Automatic Elicitation of User Preferences

Pauplin et al. [335] described the development of an interactive tool used by quality control engineers. The task is to create customised software that automatically detects and highlights defects in images of processed items. The context is a flexible manufacturing environment where changes in equipment (cameras, lighting, machinery), or in the product being created, mean that the system frequently needs reconfiguring. The concept exploited here is that interactive evolution provides a means to automatically elicit user preferences. This can replace the time-consuming, and often costly, process where a series of interviews are followed up by an image-processing expert implementing a system, which then needs iterative refinement. Instead, a good image processing system is evolved interactively. The user is shown a series of images of products containing various defects. Candidate image processing systems are applied to segment these images and the results are shown by coloured lines. Each different candidate solution will segment images in a different way, so the question asked of the user by the interface in Fig. 14.1 becomes: "Which of these sets of images has the lines drawn to separate out things you might be interested in, and how well do they do?"

[1] `http://picbreeder.org` or `http://endlessforms.com/`

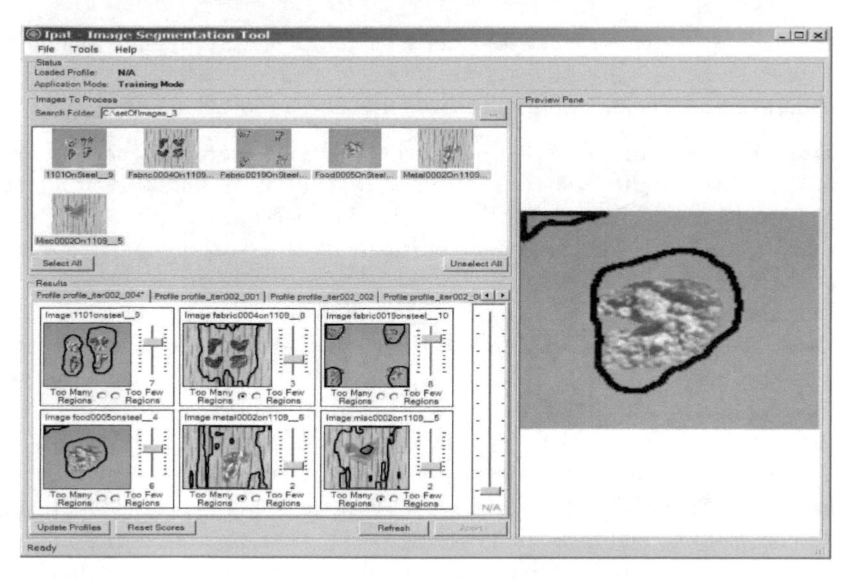

Fig. 14.1. User interface for the IPAT tool. Top window shows 'raw' images, bottom window shows segmentation results for six different images processed by current solution, right-hand side shows enlarged preview of image under cursor

In order to reduce the influence of human preconceptions, the component-based approach is used. A candidate solution, i.e., an image processing system, is composed of a variable number of image processing kernels. Each kernel consists of a module drawn from a large library of routines, together with any relevant parameter values. Kernels may be composed in a range of ways (parallel/sequential) and may be inverted to produce the final segmented version of an input image.

The algorithm, and in particular the detailed graphical user interface (GUI) design, were based on the twin principles of reducing the number and complexity of the decisions that the user had to make at each stage, and maximising the information extracted from those decisions. The user's concentration is treated as a limited resource that should be used sparingly, and is reduced by every mouse click, whereas attention is reinforced when the system appears to be working in collaboration with them. Thus, for example, the interface shown in Fig. 14.1 uses widgets that change the focus and images displayed in the right-hand 'preview' window according to mouse position rather than clicks.

Typically a session will examine several images at once, containing different types of defects, and try to find a compromise that segments them all. The interface handles this multi-objective search in two ways. Users can assign a partial fitness for each image segmented by the current solution (in which case these are averaged), or can assign an overall fitness for the solution. One mouse click allows them to switch between comparing all images segmented by a solution, or all solutions' segmentation of a single image.

Behind the interface, the principal algorithmic adaptations were:

- To reduce the number of decisions required, users were only asked to provide a score for the best solution in each generation, and only six images were presented. This necessitated a $(1 + 5)$ selection strategy.
- Only 10 different fitness levels were used to reduce the cognitive burden of scoring.
- In response to videos showing users' surprise and frustration when evolution appeared to 'forget' good solutions, one copy of the chosen solution was always shown unchanged. Comparative results showed that this evaluation regime increased the number of user interactions and the quality of the final solution (according to subjective and objective criteria).
- The applied mutation rate was determined by the users' scoring, decreasing from a high of 50% for a score of 0 to 0% for a 'perfect' solution. Results demonstrated that users' behaviour changed over time as they gained experience — so that effectively they used a low score as a reset button to counteract the high selection pressure of the $(1 + 5)$ strategy.
- Hint buttons were provided to allow the user to directly guide the direction of search. In this case domain-specific knowledge was applied to translate user input such as 'too many regions segmented' into biases in the mutation rate applied to parts of the solutions.

The most obvious component not present in this tool was the use of a surrogate fitness function. Nevertheless, the results showed that naive users were able to create image processing systems that accurately segmented a range of images, from scratch, in fewer than 20 iterations.

For exercises and recommended reading for this chapter, please visit `www.evolutionarycomputation.org`.

15

Coevolutionary Systems

In most of this book we have been concerned with problems where the quality of a proposed solution can be relatively easily measured in isolation by some externally provided fitness function. Evaluating a solution may involve an element of random noise, but does not particularly depend on the *context* in which it is done. However, there are two obvious scenarios in which this set-up does not really hold. The first occurs when a solution represents some strategy or design that works in opposition to some competitor that is itself adapting. The most obvious example here would be adversarial game-playing such as chess. The second comes about when a solution being evolved does not represent a complete solution to a problem, but instead can only be evaluated as part of a greater whole, that together accomplishes some task. An example might be the evolution of a set of traffic-light controllers, each to be sited on a different junction, with fitness reflecting their joint performance in reducing congestion over a day's simulated traffic. Both of these are examples of **coevolution**. This chapter gives an overview of the types of scenarios where coevolution might be usefully applied, and of some of the issues involved in designing a successful application.

15.1 Coevolution in Nature

Previously in this book we made extensive use of Wright's analogy of the adaptive landscape, where an evolving population is conceptualised as moving on a surface whose points represent the set of possible solutions. This metaphor ascribes a vertical dimension to the search space that denotes the fitness of a particular solution, and the combined effects of selection and variation operators move the set of points into high-fitness regions.

While an attractive metaphor, this can also be profoundly misleading when we consider the adaptation of a biological species. This is because it tends to lead to the implicit notion that solutions have a fitness value *per se*. Of course, in life the adaptive value (that is, fitness) of an organism is entirely determined by the environmental niche in which it finds itself. The characteristics of this niche are predominantly determined by the presence and character of other organisms from the same and, in particular, different species.[1]

The effect of other species in determining the fitness of an organism can be positive – for example, the pollination of plants by insects feeding on their nectar – or negative – for example, the eating of rabbits by foxes. Biologists tend to use the terms **mutualism** and **symbiosis** to refer to the coadaptation of species in a mutually beneficial way, and the terms **predation** or **parasitism** to refer to relationships in which one species has a negative effect on the survival and reproductive success of the other (antagonism).

If all of the other species in an environmental niche remained the same, and only one species was evolving, then the notion of a fixed adaptive landscape would be valid for that species. However, since evolution affects all species, the net effect is that the landscape 'seen' by each species is affected by the configuration of all the other interacting species, i.e., it will move. This process is known as coevolution. To give a concrete example, the adaptive value to a rabbit of being able to run at, say, 20 km/h depends entirely on whether the fox that preys on it has a maximum top speed of 15 km/h or 30 km/h. The height on the landscape of a 20-km/h phenotype is reduced over time from a high value to a low value as the fox evolves the ability to run faster.

Despite the additional complications of coevolutionary models, they hold some significant advantages that have been exploited within EAs to aid the generation of solutions to a range of difficult problems. One that we have already described in Sect. 6.5 is the coevolution of a population of partial models in Michigan-style LCS — these may be thought of as co-operating to provide a complete model of a problem. Another very well known example is the modelling and evolution of game-playing strategies. In this case evolving solutions play against each other to get their fitness, i.e., only one species is used and the model is competitive in the sense defined in Sect. 15.3. Since computers provide the freedom to use a number of different models, and biology is serving as an inspiration rather than a strict blueprint, a number of different models have been used successfully. Coevolutionary EAs have been implemented using both cooperation and competition, and both single and multiple-species models, as we shall now describe.

15.2 Cooperative Coevolution

Coevolutionary models in which a number of different species, each representing part of a problem, cooperate in order to solve a larger problem have been successfully applied many times. Among many examples of this are high-dimensional function optimisation [342] and job shop scheduling [225].

The advantage of this approach is that it permits effective function decomposition; each subpopulation is effectively solving a much smaller, more

[1] With the possible exception of some extremely simple organisms.

tractable problem. The disadvantage is that it relies on the user to subdivide the problem which may not be obvious from the overall specification. In nature, mutually beneficial relationships have as their ultimate expression so-called **endosymbiosis**, where the two species become so interdependent that they end up inextricably physically linked – for example, the various gut bacteria that live entirely within a host's body and are passed from mother to offspring. The equivalent in EA optimisation is where the different parts of a problem are so interdependent that they are not amenable to division.

Bull [68] conducted a series of more general studies on cooperative coevolution using Kauffman's static NKC model [244] in which the amount of effect that the species have on each other can be varied systematically. In [69] he examined the evolution of coevolving symbiotic systems that had the ability to evolve linkage flags denoting that solutions from different populations should stay together. He showed that the strategies that emerge depend heavily on the extent to which the two populations affect each other's fitness landscape, with linkage preferred in highly interdependent situations.

15.2.1 Partnering Strategies

When cooperating populations are used, a major issue is that of deciding how a solution from one population should be paired with the necessary others in order to gain a fitness evaluation.

Potter and De Jong [342] used a generational GA in each subpopulation, with the different species taking it in turns to undergo a round of selection, recombination, and mutation. Evaluation was performed using the current best from each of the other species.

Paredis coevolved solutions and their representations in a steady-state model using what he termed lifetime fitness evaluation (LTFE) [332]. In the most general form of LTFE a new individual undergoes 20 'encounters' with solutions selected from the other population. The fitness of the new individual is initially set as the mean fitness from these encounters. The effect of this scheme is that individuals from each population are continuously undergoing new encounters, and the fitness of an individual is given by the running average of its performance in the last 20 encounters. The benefit of this running-average approach is that it effectively slows down the rate at which each fitness landscape changes in response to changes in the composition of the other populations.

Husbands [225] solved the pairing problem and also effectively changed the rate at which the composition of the different populations are *perceived* to change by using a diffusion model EA (Sect. 5.5.7) with one member of each species located on each grid point.

Bull [69] examined the use of a range of different pairing strategies: best, random, stochastic fitness-based, joined, and distributed, as per [225]. His results showed that no one strategy performed better across the range of different interaction strengths and generational models, but random was robust in a generational GA, and distributed did best in a steady-state GA. When fitness sharing was added to prevent premature convergence, "best" became the most robust solution.

Finally, within the field of cooperative coevolution it is worth mentioning the use of **automatically defined functions** within GP [253]. In this extension of GP, the function set is extended to include calls to functions that are themselves being evolved in parallel, in separate populations. The great advantage of this is in permitting the evolution of modularity and code reuse.

15.3 Competitive Coevolution

In the competitive coevolution paradigm individuals compete against each other to gain fitness at each other's expense. These individuals may belong to the same or different species, in which case it is arguably more accurate to say that the different species are competing against each other.

As noted above, the classic example of this that generated much interest in the paradigm was Axelrod's work on the **iterated prisoner's dilemma** [14, 15], although early work can be traced back as far as 1962 [39]. This is a two-player game, where each participant must decide whether to cooperate or defect in each iteration. The payoff received depends on the actions of the other player, as determined by a matrix such as Table 15.1.

	Player B	
Player A	Cooperate	Defect
Cooperate	(3,3)	(0,5)
Defect	(5,0)	(1,1)

Table 15.1. Example payoff matrix for iterated prisoner's dilemma. Payoff to player A is first of pair

Axelrod organised tournaments in which human-designed strategies competed against each other, with strategies only allowed to "see" the last three actions of their opponent. He then set up experiments in which strategies were evolved using as their fitness the mean score attained against a set of eight human strategies. He was able to illustrate that the system evolved the best strategy (tit-for-tat), but there was some brittleness according to the set of human strategies chosen. In a subsequent experiment he demonstrated that a strategy similar to tit-for-tat could also be evolved if a coevolutionary approach was used with each solution playing against every other in its current generation in order to assess its quality.

In another groundbreaking study, Hillis [213] used a two-species model with the pairing strategy determined by colocation on a grid in a diffusion model EA. Note that this parallel model is similar to, and in fact was a precursor of, Husbands' cooperative algorithm described above. Hillis' two populations represented sorting networks, whose task it was to sort a number of inputs numerically, and sets of test cases for those networks. Fitness for the networks is assigned according to how many of the test cases they sort correctly. Using the antagonistic approach, fitness for the individuals representing sets of test cases is assigned according to how many errors they cause in the network's output. His study caused considerable attention as it found correct sorting networks that were smaller than any previously known.

This two-species competitive model has been used by a number of authors to coevolve classification systems [181, 333]. The approach of Paredis is worth noting as it solves the pairing strategy problem by using a variant of the LTFE method sketched above.

As with cooperative coevolution, the fitness landscapes will change as the populations evolve, and the choice of pairing strategies can have a major effect on the observed behaviour. When the competition arises within a single population, the most common approaches are to either pair each strategy against each other, or just against a randomly chosen fixed-size sample of the others. Once this has been done, the solutions can be ranked according to the number of wins they achieve and any rank-based selection mechanism chosen.

If the competition arises between different populations, then a pairing strategy must be chosen for fitness evaluation, as it is for cooperative coevolution. Since the NKC model essentially assigns random effects to the interactions between species, i.e., it is neither explicitly cooperative nor competitive, it is likely that Bull's results summarised above will also translate to this paradigm.

The main engine behind coevolution is sometimes called "competitive fitness evaluation". As Angeline states in [10], the chief advantage of the method is that it is self-scaling: early in the run relatively poor solutions may survive, for their competitors are not strong either. But as the run proceeds and the average strength of the population increases, the difficulty of the fitness function is continually scaled.

15.4 Summary of Algorithmic Adaptations for Context-Dependent Evaluation

As the discussion above shows, the choice of context, or equivalently the pairing strategy, can have a significant effect on how well EAs perform in this type of situation, and many successful approaches attempt to reduce this effect by awarding fitness as an average of a number of contexts encountered.

The second major algorithmic adaptation is the incorporation of some kind of history into the evaluation — often in the form of an archive of historically 'good solutions against which evolving solutions are periodically tested. The reason for this is to avoid the problem of cycling: a phenomenon where evolution repeatedly moves through a series of solutions rather than making advances. To illustrate this consider as an example the simple 'rock–scissors–paper game. In a population which had converged to a 'rock strategy, a mutant individual displaying 'scissors would be disadvantaged, whereas a 'paper strategy would do well and come to dominate the population. This in turn provides the conditions to favour a chance mutation into a 'scissors strategy, and later back to 'rock', and so on. This form of cycling can be avoided by the use of an archive of different solutions, which can provide impetus for the evolution of more complex strategies, and has been demonstrated in a range of applications.

15.5 Example Application: Coevolving Checkers Players

In [81], which is expanded into a highly readable book [168] and further summarised in [169], Fogel charts the development of a program for playing the game of checkers (a.k.a. draughts), a board game that is also highly popular on the Internet. In this two-player game a standard 8×8 squared board is used, and each player has an (initially fixed) number of pieces (checkers), which move diagonally on the board. A checker can 'take' an opponent's piece if it is adjacent, and the checker jumps over it into an empty square (both players use the same-coloured squares on the board). If a checker reaches the opponent's home side, it becomes a 'king' in which case it can move backwards as well as forwards. Human checker players regularly compete against each other in a variety of tournaments (often Internet-hosted), and there is a standard scheme for rating a player according to their results.

In order to play the game, the program evaluates the future value of possible moves. It does this by calculating the likely board state if that move is made, using an iterative approach that looks a given distance ('ply') into the future. A board state is assigned a value by a neural network, whose output is taken as the 'worth' of the state from the perspective of the last player to move.

The board state is presented to the neural network as a vector of length 32, since there are 32 possible board sites. Each component comes from the set $\{-K, -1, 0, 1, K\}$, where the minus sign presents an opponent's king or piece, and K takes a value in the range [1.0, 3.0].

The neural network thus defines a "strategy" for playing the game, and this strategy is evolved with evolutionary programming. A fixed structure is used for the neural networks, which has a total of 5046 weights and bias terms that are evolved, along with the importance given to the kings K. An individual solution is thus a vector of dimension 5047.

The authors used a population size of 15, with a tournament size 5. When programs played against each other they scored +1, 0, −2 points for a win, draw, and loss, respectively. The 30 solutions were ranked according to their scores over the 5 games, then the best 15 became the next generation.

The mutation operator used took two forms: the weights/biases were mutated using the addition of Gaussian noise, with lognormal adaptation of the step sizes *before* mutation of the variables, i.e., using standard self-adaptation with $n = 5046$ strategy parameters. The offspring king weightings were mutated accorded to $K' = K + \delta$, where δ is sampled uniformly from $[-0.1, 0.1]$, and the resulting values of K' are constrained to the range $[1.0, 3.0]$. Weights and biases were initialised randomly over the range $[-0.2, 0.2]$. K values were initially set to 2.0, and the strategy parameters were initialised to 0.05.

The authors proceeded by having the neural networks compete against each other for 840 generations (6 months) before taking the best evolved strategy and testing it against human opponents on the Internet. The results were highly impressive: over a series of trials the program earned an average ranking that put it in the "expert" class, and better than 99.61% of all rated players on the website. This work is particularly interesting in the context of artificial intelligence research for the following reasons:

- There is no input of human expertise about good short-term strategies or endgames.
- There is no input to tell the evolving programs that in evaluating board positions a negative vector sum (that is, the opponent has a higher piece-count) is worse than a positive vector sum.
- There is no explicit credit assignment'mechanism to reward moves that lead to wins; rather a 'top-down' approach is taken that gives a single reward for an entire game.
- The selection function averages over five games, so the effects of strategies that lead to wins or losses are blurred.
- The strategies evolve by playing against themselves, with no need for human intervention!

In the spirit of this chapter, we can say that the approach taken is rather similar to Paredis' Life-time fitness evaluation — to even out the effect of a specific context (combination of opponents) fitness is awarded as the average performance across a number of contexts.

For exercises and recommended reading for this chapter, please visit `www.evolutionarycomputation.org`.

16

Theory

In this chapter we present a brief overview of some of the approaches taken to analysing and modelling the behaviour of evolutionary algorithms. The Holy Grail of these efforts is the formulation of predictive models describing the behaviour of an EA on arbitrary problems, and permitting the specification of the most efficient form of optimiser for any given problem. However, (at least in the authors' opinions) this is unlikely ever to be realised, and most researchers will currently happily settle for techniques that provide *any* verifiable insights into EA behaviour, even on simple test problems. The reason for what might seem like limited ambition lies in one simple fact: evolutionary algorithms are hugely complex systems, involving many random factors. Moreover, while the field of EAs is fairly young, it is worth noting that the field of population genetics and evolutionary theory has a head start of more than a hundred years, and is still battling against the barrier of complexity.

Full descriptions and analysis of the various techniques currently used to develop EA theory would require both an amount of space and an assumption of prior knowledge of mathematics and statistics that are unsuitable here. We therefore restrict ourselves to a fairly brief description of the principal methods and results which historically informed the field. We begin by describing some of the approaches taken to modelling EAs using a discrete representation (i.e., for combinatorial optimisation problems), before moving on to describe the techniques used for continuous representations. This chapter finishes with a description of an important theoretical result concerning all optimisation algorithms, the No Free Lunch (NFL) theorem.

For further details, we point the interested reader to 'bird's eye overviews' such as [140], and extensive monographs such as [52, 446, 353]. For a good overview of the most promising recent approaches and results we would suggest [234], or collections such as [63].

16.1 Competing Hyperplanes in Binary Spaces: The Schema Theorem

What Is a Schema?

Since Holland's initial analysis, two related concepts have dominated much of the theoretical analysis and thinking about GAs. These are the concepts of **schema** (plural schemata) and **building blocks**. A schema is simply a hyperplane in the search space, and the common representation of these for binary alphabets uses a third symbol — # the "don't care" symbol. Thus for a five-bit problem, the schema 11### is the hyperplane defined by having ones in its first two positions. All strings meeting this criterion are **instances**, or examples, of this schema (in this case there are $2^3 = 8$ of them). The fitness of a schema is the mean fitness of all strings that are examples of it; in practice this is often estimated from samples when there are many such strings. Global optimisation can be seen as the search for the schema with zero "don't care" symbols, which has the highest fitness.

Holland's initial work showed that the analysis of GA behaviour was far simpler if carried out in terms of schemata. This is an example of **aggregation** in which rather than model the evolution of all possible strings, they are grouped together in some way and the evolution of the aggregated variables is modelled. He showed that a string of length l is an example of 2^l schemata. Although in general there will not be as many as $\mu \cdot 2^l$ distinct schemata in a population of size μ, he derived an estimate that a population will usefully process $O(\mu^3)$ schemata. This result, known as **implicit parallelism** is widely quoted as being one of the main factors in the success of GAs.

Two features are used to describe schemata. The **order** is the number of positions in the schema that do not have the # sign. The **defining length** is the distance between the outermost defined positions (which equals the number of possible crossover points between them). Thus the schema H=1##0#1#0 has order $o(H) = 4$ and defining length $d(H) = 8 - 1 = 7$.

The number of examples of a schema in an evolving population depends on the effects of variation operators. While selection operators can only change the relative frequency of pre-existing examples, operators such as recombination and mutation can both create new examples and disrupt current ones. In what follows we use the notation $Pd(H, x)$ to denote the probability that the action of an operator x on an instance of a schema H is to destroy it, and $Ps(H)$ to denote the probability that a string containing an instance of schema H is selected.

Holland's Formulation for the SGA

Holland's analysis applied to the standard genetic algorithm (SGA) using fitness proportionate parent selection, one-point crossover (1X), and bitwise mutation, with a generational survivor selection. Considering a genotype of length l that contains an example of a schema H, the schema may be disrupted if the crossover point falls between the ends, which happens with probability

$$Pd(H, 1X) = \frac{d(H)}{(l-1)}.$$

The chance that bitwise mutation with rate P_m will disrupt the schema H is proportional to the order of the schema: $Pd(H, mutation) = 1 - (1 - P_m)^{o(H)}$. After expansion, and ignoring high-order terms in P_m, this approximates to

$$Pd(H, mutation) = o(H) \cdot P_m.$$

The probability of a schema being selected depends on the fitness of the individuals in which it appears relative to the total population fitness, and the number of examples present $n(H,t)$. Using $f(H)$ to represent the **fitness of the schema** H, defined as the mean fitness of individuals that are examples of schema H, and $<f>$ to denote the mean population fitness, we obtain:

$$Ps(H,t) = \frac{n(H,t) \cdot f(H)}{\mu \cdot <f>}.$$

μ independent samples are taken to create the next set of parents, so the expected number of instances of H in the population after selection is:

$$n'(H,t) = \mu \cdot Ps(H,t) = \frac{n(H,t) \cdot f(H)}{<f>}.$$

After normalising by μ (to make the result population-size independent), allowing for the disruptive effects of recombination and mutation derived above, and using an inequality to allow for the creation of new instances of H by the variation operators, the proportion $m(H)$ of individuals representing schema H at subsequent time-steps is given by:

$$m(H, t+1) \geq m(H,t) \cdot \frac{f(H)}{<f>} \cdot \left[1 - \left(p_c \cdot \frac{d(H)}{l-1}\right)\right] \cdot [1 - p_m \cdot o(H)], \quad (16.1)$$

where p_c and p_m are the probabilities of applying crossover, and the bitwise mutation probability, respectively.

This is the **schema theorem**, and the original understanding of this result was that schemata of above-average fitness would increase their number of instances within the population from generation to generation. We can quantify this by noting that the condition for a schema to increase its representation is $m(H, t+1) > m(H,t)$ which is equivalent to:

$$\frac{f(H)}{<f>} > \left[1 - \left(p_c \cdot \frac{d(H)}{l-1}\right)\right] \cdot [1 - p_m \cdot o(H)].$$

Schema-Based Analysis of Variation Operators

Holland's original version of the schema theorem, as formulated above, was for one-point crossover and bitwise mutation. Following the rapid proliferation of alternative variation (particularly recombination) operators as the field expanded and diversified, a considerable body of results was developed to try and understand why some operators gave improved performance on certain problems. Particularly worthy of mention within this were two long-term research programs. Over a number of years, Spears and De Jong developed analytical results for $Pd(H, x)$ as a function of defining length $d(H)$ and order $o(H)$ for a number of different recombination and mutation operators [107, 108, 408, 412, 413, 414], which are brought together in [411].

Meanwhile, Eshelman and Schaffer conducted a series of empirical studies [157, 159, 161, 366] in which they compared the effects of mutation with various crossover operators on the performance of a GA. They introduced the notion of **operator bias** to describe the interdependence of $Pd(H, x)$ on $d(H), o(H)$ and x, which takes two forms:

- If an operator x displays **positional bias** it is more likely to keep together bits that are close together in the representation. This has the effect that given two schemata H_1, H_2 with $f(H_1) = f(H_2)$ and $d(H_1) < d(H_2)$, then $Pd(H_1, x) < Pd(H_2, x)$.
- By contrast, if an operator displays **distributional bias** then the probability that it will transmit a schema is a function of $o(H)$. One example of this is bitwise mutation, where, as we have seen, the probability of disruption increases with the order: $Pd(H, mutation) \approx Pm \cdot o(H)$. Another example is uniform crossover which will on average select half of the genes from one parent, and so is increasingly likely to disrupt a schema as the ratio $o(H)/l$ increases beyond 0.5.

Although these results provided valuable insight and have informed many practical implementations, it is worth bearing in mind that they are only considering the disruptive effects of operators. Analysis of the constructive effects of operators in creating new instances of a schema H are harder, since these effects depend heavily on the constitution of the current population. However, under some simplifying assumptions, Spears and De Jong [414] developed the surprising result that the expected number of instances of a schema destroyed by a recombination operator is equal to the expected number of instances created, for all recombination operators!

Walsh Analysis and Deception

If we return our attention to the derivation of the schema theorem, we can immediately see from an examination of the disruption probabilities given above that *all other things being equal*, short low-order schema have a greater chance of being transmitted to the next generation than longer or higher-order schema of the same mean fitness. This analysis has led to what has become known as the **building block hypothesis** [189, pp. 41–45]: that GAs begin by selecting amongst competing short low-order schemata, and then progressively combine them to create higher-order schemata, repeating this process until (hopefully) a schema of length $l-1$ and order l, i.e., the globally optimal string, is created and selected for. Note that for two schemata to compete they must have fixed bits (1 or 0) in the same positions. Thinking along these lines raised the obvious question: "What happens if the global optimum is *not* an example of the low-order schemata that have the highest mean fitness?".

To give an immediate example, let us consider a four-bit problem that has 0000 as its global optimum. It turns out that it is relatively simple to create the situation where all of the order-n schemata containing 0's in their defining positions are less fit than the corresponding schemata with 1's in those position, i.e., $f(0\#\#\#) < f(1\#\#\#)$, $f(\#0\#\#) < f(\#1\#\#)$, etc., right up to $f(\#000) < f(\#111)$, $f(0\#00) < f(1\#11)$, etc. All that is required to achieve this is that the fitness of a globally optimal string is sufficiently greater than all the other strings in every schema of which it is a member. In this case we might expect that every time the GA makes a decision between two order-n schemata, it is likely to make the wrong decision unless n=4.

This type of problem is known as **deceptive** and has been of great interest since it would appear to make life hard for a GA, in that the necessary building blocks for successful optimisation are not present. However, it has been postulated that if a fitness function is composed of a number of deceptive problems, then at least a GA using recombination offers the possibility that these can be solved independently and mixed via crossover. By comparison, an optimisation technique relying on local search continuously makes decisions on the basis of low-order schema, and so is far more likely to be 'fooled'. Note that we have not provided a formal definition of the conditions necessary for a function to be deceptive; much work has been done on this subject and slightly differing definitions exist [200, 403, 455].

The importance of deceptive problems to GA theory and analysis is debatable. At various stages some eminent practitioners have made claims that "the only challenging problems are deceptive" [93], (although this view may have been modified with hindsight), but others have argued forcibly against the relevance of deception. Grefenstette showed that it is simple to circumnavigate the problem of deception in GAs by looking for the best solution in each new generation and then creating its inverse [200]. Moreover, Smith and Smith created an abstract randomised test problem generator (NKPRS) in which the probability that a landscape was deceptive could be directly manipulated [404]. Their findings did not demonstrate that there was a correlation between the likelihood of deception and the ability of a standard GA to discover the global optimum.

Much of the work in this area makes use of **Walsh functions** to analyse fitnesses. This technique was first used for GA analysis in [51], but became more widely known after a series of important papers by Goldberg [187, 188]. These are a set of functions that provide a natural basis for the decomposition of a binary search landscape. They can be thought of as equivalent to the way that Fourier transforms decompose a complex signal in the time domain into a weighted sum of sinusoidal waves, which can be represented and manipulated in the frequency domain. Just as Fourier transforms form a vital part in a huge range of signal processing and other engineering applications, because sine functions are so easily manipulable, so Walsh transforms form an easily manipulable way of analysing binary search landscapes, with the added bonus that there is a natural correspondence between Walsh partitions (the equivalent of harmonic frequencies) and schemata. For more details on Walsh analysis the reader is directed to [187] or [353].

16.2 Criticisms and Recent Extensions of the Schema Theorem

Despite the attractiveness of the schema theorem as a description for how GAs work, it has come in for a good amount of criticism, and significant quantities of experimental evidence and theoretical arguments have been produced to dispute its importance. This is perhaps inevitable given that early on some rather extravagant claims were made by its adherents, and given the perhaps natural tendency of humans to take pot-shots at 'sacred cows'.

Ironically, empirical counterevidence was provided by Holland himself, in conjunction with Mitchell and Forrest, who created the **Royal Road functions** based on schema ideas in order to demonstrate the superiority of GAs over local search methods. Unfortunately, their results demonstrated that the opposite was in fact true [177]! However, this work did lead to the understanding of the phenomenon of **hitchhiking** whereby an unfavourable allele becomes established in the population because of an early association with an instance of a high-fitness schema.

Theoretical arguments against the value of the schema theorem and associated analysis have included:

- Even if it is correctly estimated, the rate of increase in representation of any given schema is not in fact exponential. This is because its selective advantage $f(H)/ <f>$ decreases as its share of the population increases and the mean fitness rises accordingly.
- Eq. (16.1) applies to the *estimated* fitness of a given schema as averaged over all the instances in the current population, which might not be representative of the schema as a whole. Thus although the schema theorem is correct in predicting the frequency of a schema in the next generation, it can tell us almost *nothing* about the frequency in future generations, since

as the proportions of other schema change, so will the composition of the set of strings which represent H, and hence the estimates of $f(H)$.

- Findings that Holland's idea that fitness proportionate selection allocated optimal amounts of trials to competing schemata is incorrect [277, 359].
- The fact that the schema theorem ignores the constructive effects of operators. Altenberg [6] showed that in fact the schema theorem is a special case of Price's theorem in population genetics. This latter includes both constructive and disruptive terms. Whilst exact versions of the schema theorem have recently been derived [418], these currently remain somewhat intractable even for relatively simple test problems, although their use is starting to offer interesting new perspectives.

These arguments and more are summarised eloquently in [353, pp. 74–90]. We should point out that despite these criticisms, schemata represent a useful tool for understanding *some* of how GAs work, and we would wish to stress the vital role that Holland's insights into the importance of schemata have had in the development of genetic algorithms.

16.3 Gene Linkage: Identifying and Recombining Building Blocks

The Building Block Hypothesis offers an explanation of the operation of GAs as a process of discovering and putting together blocks of coadapted genes of increasing higher orders. To do this, it is necessary for the GA to discriminate between competing schemata on the basis of their estimated fitness. The Messy GA [191] was an attempt to explicitly construct an algorithm that worked in this fashion. The use of a representation that allowed variable length strings and removed the need to manipulate strings in the order of their expression began a focus on the notion of gene linkage (in this context gene is taken to mean the combination of a particular allele value and locus).

Munetomo and Goldberg [312] identify three approaches to the identification of linkage groups. The first of these they refer to as the "direct detection of bias in probability distributions", and is exemplified by Estimation of Distribution Algorithms described in Section 6.8. Common to all of these approaches is the notion of first identifying a factorisation of the problem into a number of subgroups, such that a given statistical criterion is minimised, based on the current population. This corresponds to learning a linkage model of the problem. Once these models have been derived, conditional probabilities of gene frequencies within the linkage groups are calculated, and a new population is generated based on these, replacing the traditional recombination and mutation steps of an EA. It should be emphasised that these EDA approaches are based on statistical modelling rather than on a schema-based analysis. However, since they implicitly construct a linkage analysis of the problem, it would be inappropriate not to mention them here.

The other two approaches identified by Munetomo and Goldberg use more traditional recombination and mutation stages, but bias the recombination operator to use linkage information.

In [243, 312] first-order statistics based on pairwise perturbation of allele values are used to identify the blocks of linked genes that algorithms manipulate. Similar statistics are used in a number of other schemes such as [438].

The third approach identified does not calculate statistics on the gene interactions based on perturbations, but rather adapts linkage groups explicitly or implicitly via the adaptation of recombination operators. Examples of this approach that explicitly adapt linkage models can be seen in [209, 362, 393, 394, 395]. A mathematical model of the linkage models of different operators, together with an investigation of how the adaptation of linkage must happen at an appropriate level (see Sect. 8.3.4 for a discussion of the issue of the scope of adaptation), can be found in [385].

16.4 Dynamical Systems

The **dynamical systems** approach to modelling EAs in finite search spaces has principally been concerned with genetic algorithms because of their (relative) simplicity. Michael Voses established the basic formalisms and results in a string of papers culminating in the publication of his book [446]. This work has been taken on and extended by a number of authors (see, for example, the proceedings of the Foundations of Genetic Algorithms workshops [38, 285, 341]). The approach can be characterised as follows:

- Start with an n-dimensional vector $\overline{p}$, where n is the size of the search space, and the component p_i^t represents the proportion of the population that is of type i at iteration t.
- Construct a **mixing matrix** M representing the effects of recombination and mutation, and a **selection matrix** F representing the effects of the selection operator on each string for a given fitness function.
- Compose a genetic operator $G = F \circ M$ as the matrix product of these two functions.
- The action of the GA to generate the next population can then be characterised as the application of this operator G to the current population: $\overline{p}^{t+1} = G\overline{p}^t$.

Under this scheme the population can be envisaged as a point on what is known as the simplex: a surface in n-dimensional space made up of all the possible vectors whose components sum to 1.0 and are nonnegative. The form of G governs the way that a population will trace a trajectory on this surface as it evolves. A common way of visualising this approach is to think of G as defining a 'force-field' over the simplex describing the direction and intensity of the forces of evolution acting on a population. The form of G alone determines which points on the surface act as **attractors** towards which the population is drawn; and analytical analysis of G, and its constituents F and M, has led to many insights into GA behaviour.

Vose and Liepens [447] presented models for F and M under fitness proportionate selection, one-point crossover and bitwise mutation, and these have been extended to other operators in [446]. One of the insights gained by analysing the form of M is that schemata provided a natural way of aggregating strings into equivalence classes under recombination and mutation, which provides a nice tie-in to Holland's ideas.

Other authors have examined a number of alternative ways of aggregating the elements in the search space into a smaller number of equivalence classes, so as to make the models more amenable to solution. Using this approach, a number of important results have been derived, explaining facets of behaviour such as the punctuated equilibria effect (described qualitatively in [447] but expanded and including for the first time *accurate* predictions of the time spent between the discovery of new fitness levels in [439]). These ideas have also been applied to model mechanisms such as self-adaptive mutation in binary coded GAs [383, 386].

It is worth pointing out that while this model exactly predicts the *expected* proportions of different individuals present in evolving populations, these values can only be attained if the population size is infinite. For this reason this approach falls into a class known as **infinite population models**. For finite populations, the evolving vectors $\overline{p}$ can be thought of as representing the probability distribution from which μ independent samples are drawn to generate the next population. Because the smallest proportion that can be present in a real population has a size $1/\mu$, this effectively constrains the population to move between a subset of points on the simplex representing a lattice of size $1/\mu$. This means that, given an initial population, the trajectory predicted may not actually be attainable, and corrections must be made for finite population effects. This work is still ongoing.

16.5 Markov Chain Analysis

Markov chain analysis is a well-established technique that is used to study the properties and behaviour of stochastic systems. A good description can be found in many textbooks on stochastic processes [216]. For our purposes it is sufficient to note that we can describe a system as a discrete-time **Markov chain** provided that the following conditions are met:

- At any given time the system is in one of a finite number (N) of states.
- The probability that the system will be in any given state X^{t+1} in the next iteration is solely determined by the state that it is in at the current iteration X^t, regardless of the previous sequence of states.

The impact of the second condition is that we can define a **transition matrix** Q where the entry Q_{ij} contains the probability of moving from state

i to state j in a single step ($i, j \in \{1, \ldots, N\}$). It is simple to show that the probability that after n steps the system has moved from state i to state j is given by the (i, j)th entry of matrix Q^n. A number of well-known theorems and proofs exist for making predictions of the behaviour of Markov chains.

There are a finite number of ways in which we can select a finite sized population from a finite search space, so we can treat any EA working within such a representation as a Markov chain whose states represent the different possible populations, and a number of authors have used these techniques to study evolutionary algorithms.

As early as in 1989 Eiben et al. [1, 129] proposed a Markov model for the abstract genetic algorithm built from a choice, a production, and a selection function, and used it to establish convergence properties. In contemporary terminology it is a general framework for EAs based on parent selection, variation, and survivor selection, respectively. It has been proved that an EA optimising a function over an arbitrary finite space converges to an optimum with probability 1 under some rather permissive conditions. Simplifying and reformulating the results, it is shown that if, in any given population,

- every individual has a nonzero probability of selection as a parent, and
- every individual has a nonzero probability of selection as a survivor, and
- the survival selection mechanism is elitist, and
- any solution can be created by the action of variation operators with a nonzero probability,

then the nth generation certainly contains the global optimum for some n.

Rudolph [357] tightened the assumptions and showed that a genetic algorithm with nonzero mutation and elitism will always converge to the globally optimal solution, but that this would not necessarily happen if elitism was not used. In [358] the convergence theorems are extended to EAs working in arbitrary (e.g., continuous) search spaces.

A number of authors have proposed exact formulations for the transition matrices Q of binary coded genetic algorithms with fitness proportionate selection, one-point crossover, and bit-flipping mutation [99, 321]. They essentially work by decomposing the action of a GA into two functions, one of which encompasses recombination and mutation (and is purely a function of the crossover probability and mutation rate), and the other that represents the action of the selection operator (which encompasses information about the fitness function). These represent a significant step towards developing a general theory; however, their usefulness is limited by the fact that the associated transition matrices are enormous: for an l-bit problem there are $\binom{\mu + 2^l - 1}{2^l - 1}$ possible populations of size μ and this many rows and columns in the transition matrix.

It is left as an exercise for the reader to calculate the size of the transition matrix for a ten-bit problem with ten members in the population, in order to get a feel for how likely it is that advances in computing will make it possible to manipulate these matrices.

16.6 Statistical Mechanics Approaches

The **statistical mechanics** approach to modelling EA behaviour was inspired by the way that complex systems consisting of ensembles of many smaller parts have been modelled in physics. Rather than trying to trace the behaviour of all the elements of a system (the **microscopic** approach), this approach focuses on modelling the behaviour of a few variables that characterise the system. This is known as the **macroscopic approach**. There are obvious links to the aggregating versions of the dynamical systems approach described above; however, the quantities modelled are related to the cumulants of the variables of interest [345, 346, 348, 354].

Thus if we are interested in the fitness of an evolving population, equations are derived that yield the progress of the moments of fitness $<f>, <f^2>$, $<f^3>$, and so on (where the braces $<>$ denote that the mean is taken over the set of possible populations) under the effects of selection and variation. From these properties, cumulants such as the mean ($<f>$ by definition), variance, skewness, etc., of the evolving population can be predicted as a function of time. Note that these predictions are necessarily approximations whose accuracy depends on the number of moments modelled.

The equations derived rely on various 'tricks' from the statistical mechanics literature and are predominantly for a particular form of selection (Boltzmann selection). The approach does not pretend to offer predictions other than of the population mean, variance and so on, so it cannot be used for all the aspects of behaviour one might desire to model. These techniques are nevertheless impressively accurate at predicting the behaviour of real GAs on a variety of simple test functions. In [347] Prügel-Bennett compares this approach with a dynamical systems approach based on aggregating fitness classes and concludes that the latter approach is less accurate at predicting dynamic behaviour of the population mean fitness (as opposed to the long-term limit) because the variables that it tracks are not representative as a result of the averaging process. Clearly this work deserves further study.

16.7 Reductionist Approaches

So far we have described a number of methods for modelling the behaviour of EAs that attempt to make predictions about the composition of the next population by considering the effect of all the genetic operators on the current population. We could describe these as holistic approaches, since they explicitly recognise that there will be interactions between the effects of different operators on the evolving population. An unfortunate side effect of this holistic approach is that either the resulting systems become very difficult to manipulate, as a result of their sheer size, or necessarily involve approximations and may not model all of the variables that we would like to predict.

An alternative methodology is to take a reductionist approach, and examine parts of the system separately. Although ultimately flawed in neglecting interaction effects, this approach is common to many branches of physics and engineering, where it has been used to yield frequently accurate predictions and insights, provided that a suitable decomposition of the system is made.

The advantage of taking a reductionist approach is that frequently it is possible to derive analytical results and insights when only a part of the problem is considered. A typical division is between selection and variation. A great deal of work has been done on characterising the effects of different selection operators, which can be thought of as complementary to the work described in Section 16.1.

Goldberg and Deb [190] introduced the concept of **takeover time**, which is the number of generations needed for a single copy of the fittest string to completely take over the population in a "selecto-EA" (i.e., one in which no variation operators are used). This work has been extended to cover a variety of different mechanisms for parental and survivor selection, using a variety of theoretical tools such as difference equations, order statistics, and Markov chains [19, 20, 21, 58, 78, 79, 360, 400].

Parallel to this, Goldberg, Thierens, and others examined what they called the **mixing time**, which characterises the speed at which recombination brings together building blocks initially present in different members of a population [430]. Their essential insight is that in order to build a well-performing EA, in particular a GA, it is necessary for the mixing time to be less than the takeover time, so that all possible combinations of the building blocks present can be tried before one fitter string takes over the population and removes some of them. While the rigour of this approach can be debated, it does have the immense benefit of providing practical guidelines for population sizing, operator probabilities, choice of selection methods, and so on, which can be used to help design an effective EA for new applications.

16.8 Black Box Analsyis

One of the approaches which has yielded the most promising advances since the first edition of this book was written has been the 'black box complexity' approach introduced by Droste, Jansen and Wegener [120]. A good recent review can be found in [234], or in collections such as [63]. The essence of this approach is to model the run-time complexity of algorithms on specific functions – that is to say on their expected time from an arbitrary starting point to reaching the global optima. This is done by modelling the process as a system of steps whose likelihood can be expressed and then deriving upper and lower bounds on the run-time from these equations. This approach has lead to many useful insights on the behaviour of population-based methods, and has in part settled some long-running debates within the field – for example, by illustrating non-artificial problems on which crossover is provably useful [118].

16.9 Analysing EAs in Continuous Search Spaces

In contrast to the situation with discrete search spaces, the state of theory for continuous search spaces, and evolution strategies in particular, is fairly advanced. As noted in Section 16.5, Rudolph has shown the existence of global proofs of convergence also in such spaces [358], since the evolution of the population is itself a Markov process. Unfortunately, it turns out that the Chapman–Kolmogorov equation describing this is intractable, so the population probability distribution as a function of time cannot be determined directly. However, it turns out that much of the dynamics of ESs can be recovered from simpler models concerning the evolution of two macroscopic variables, and many theoretical results have been obtained on this basis.

The first of the variables modelled is the **progress rate**, which measures the distance of the centre of mass of the population from the global optimum (in variable space) as a function of time. The second is the **quality gain**, which measures the expected improvement in fitness between generations.

Most of this analysis has concerned variants of two fitness functions, the **sphere model**: $f(\overline{x}) = \sum_i x_i^n$ for some n, and the **corridor model** [373]. The latter takes various forms but essentially contains a single direction in which fitness is improving, hence the name. Since an arbitrary fitness function in a continuous space can usually be expanded (using a Taylor expansion) to a sum of simpler terms, the vicinity of a local optimum of one of these models is often a good approximation to the *local* landscape.

The continuous nature of the search space, coupled with the use of normally distributed mutations and well-known results from order statistics, have permitted a *relatively* straightforward derivation of equations describing the motion of the two macroscopic variables over time as a function of the values of μ, λ, and σ, starting with Rechenberg's analysis of the (1+1) ES on the sphere model, from which he derived the 1/5 success rule [352]. Following from this, the principles of self-adaptation and multimembered strategies have also been analysed. A thorough overview of these results is given in [53].

16.10 No Free Lunch Theorem

By now we hope the reader will have realised that the search for a mathematical model of EAs, which will permit us to make accurate predictions of a given algorithm on any given problem, is still a daunting distance from its goal. Whilst the tools are now in place to make some accurate predictions of some aspects of behaviour on some problems, these are often restricted to simple problems for which an EA is almost certainly not the most efficient algorithm anyway.

However, a recent line of work has come up with a result that allows us to make some statements about the comparative performance of different algorithms across all problems: they are all the same! This result is known as the **No Free Lunch theorem** (NFL) [467]. In layperson's terms it says that if we average over the space of all possible problems, then all nonrevisiting **black box algorithms** will exhibit the same performance.

By nonrevisiting we mean that the algorithm does not generate and test the same point in the search space twice. Although not typically a feature of EAs, this can simply be achieved by implementing an archive of all solutions ever seen, and then each time we generate an offspring discarding it and repeating the process if it already exists in the archive. An alternative approach (taken by Wolpert and Macready in their analysis) is to view performance as the number of distinct calls to the evaluation function. In this case we still need an archive, but we can allow duplicates in the population. By black box algorithms we mean those that do not incorporate any problem or instance-specific knowledge.

There has been some considerable debate about the utility of the No Free Lunch theorem, often centred around the question of whether the set of problems that we are likely to try to tackle with EAs is representative of all problems, or forms some special subset. However, they have come to be widely accepted, and the following lessons can be drawn:

- If we invent a new algorithm and it appears to be the best ever at solving some particular class of problems, then it will pay for this by performing poorly at some others. This suggests that a careful strategy is required to evaluate new operators and algorithms, as discussed in Chap. 9.
- *For a given problem* we can circumvent the NFL theorem by incorporating problem-specific knowledge. This of course leads us towards memetic algorithms (cf. Chap. 10).

For exercises and recommended reading for this chapter, please visit `www.evolutionarycomputation.org`.

17

Evolutionary Robotics

In this chapter we discuss evolutionary robotics (ER), where evolutionary algorithms are employed to design robots. Our emphasis lies on the evolutionary aspects, not on robotics per se. Therefore, we only briefly discuss the ER approaches that work with conventional evolutionary algorithms to optimize some robotic features and pay more attention to systems that can give rise to a new kind of evolutionary algorithms. In particular, we consider groups of mobile robots whose features evolve in real-time, for example, a swarm of Mars explorers or 'robot moles' mining ore deep under the surface. In such settings the group of robots is a population itself, which leads to interesting interactions between the robotic and the evolutionary components of the whole system. For robotics, this new kind of ER offers the ability to evolve controllers as well as morphology in partially unknown and changing environments on the fly. For evolutionary computing, autonomous mobile robots provide a special substrate for implementing and studying artificial evolutionary processes in physical entities going beyond the digital systems of today's evolutionary computing.

17.1 What Is It All About?

Evolutionary robotics is part of a greater problem domain that does not have a precise definition, but it can be characterised as problems involving physical environments. (The first edition of this book treated it very briefly in Sect. 6.10.) Evolutionary systems in this domain have populations whose members are not just points in some abstract, stateless search space, e.g., the set of all permutations of $1, \ldots, n$, but are embedded in real time and real space. In other words, such systems feature 'situated evolution' or 'physically embodied evolution' as discussed in [371]. Examples include robot swarms where the robot controllers are evolving on-the-fly, artificial life systems where predators and prey coevolve in a simulated world[1], or adaptive control systems to regulate a chemical factory where the adaptive force that changes the control policy over time is an evolutionary algorithm. In all these examples the entities undergoing evolution are *active*, in the sense that they do something, i.e., they exhibit some behaviour that can change the environment. This distinguishes them from applications evolving *passive* entities, e.g., using evolutionary algorithms for combinatorial or numerical optimisation. Hence, one of the main distinguishing features of problems in this area is that the algorithmic goal is formulated in terms of functional, rather than structural properties. In other words, it is some exhibited behaviour we are after.

The fundamental problem in the design of autonomous robots is that the targeted robot behaviour is a "dynamical process resulting from nonlinear interactions between the robot's control system, its body, and the environment" [163]. Therefore, the robot's makeup, which falls under the designer's influence, has only a partial and indirect effect on the desired behaviour. This implies that the link between the controllable parameters and the target variables is weak, ill-defined, and noisy. Furthermore, the solution space can be – and typically is – subject to conflicting constraints and objectives. Think, for instance, of trying to be fast while also sparing the batteries. Evolutionary algorithms offer a very promising approach here, because they do not require a crisply defined fitness function, they can cope with constraints and multiple objectives, and they can find good solutions in complex spaces even under noisy and dynamically changing conditions, as discussed in the previous chapters.

The use of evolutionary computing techniques to solve problems arising in robotics began in the 1990s and now evolutionary robotics has a large body of related work [322, 432, 450]. Reviewing the field is beyond the scope of this chapter, the surveys in [61, 163, 449] give a good impression of the area. The picture that arises from the literature shows a large variety of work with different robot morphologies, different controller architectures, and different tasks in different environments, including underwater and flying robots. Some of the evolved solutions are really outstanding, but many roboticists are still skeptical and the evolutionary approach to robot design is outside of the mainstream. As phrased by Bongard in [61]: "The evolution of robot bodies and brains differs markedly from all other approaches to robotics".

17.2 Introductory Example

As an introductory example let us consider the problem of evolving a controller for a wheeled robot that must ride towards the light source in a flat and empty arena. Regarding the hardware we can assume a simple cylindric robot

[1] Note that we consider physical environments in a broad sense, including simulated environments, as long as they include some notion of space and time.

with a couple of wheels and a few LED lights, a camera, a gyroscope, and four infrared sensors, one on each side. As for the control software, the robot can use a neural network that receives inputs from the **sensors** (camera, infrared sensors, gyroscope) and gives instructions to the **actuators** (motors and LEDs). Thus, a **robot controller** is a mapping between sensory inputs and actuatory outputs. Its quality – which determines its fitness used in an evolutionary algorithm – depends on the behaviour it induces. For instance, the time needed from start to hitting the source can be used for this purpose.

As with any application of an EA, an essential design decision when evolving robot controllers is to distinguish phenotypes and genotypes, cf. Chap. 3. Simply put, this distinction means that we perceive the controllers with all their structural and procedural complexity as phenotypes and we introduce a structurally more simple representation of the controllers as genotypes. Obviously, we also need a mapping from genotypes to phenotypes, which might be a simple mapping, or a complex transformation. For example, a robot controller may consist of a collection of artificial neural nets (ANNs) and a decision tree, where the decision tree specifies which ANN will be invoked to produce the robot's response in a given situation. This decision tree can be as simple as calling ANN-1 when the room is lit and calling ANN-2 when the room is dark. This complex controller, i.e., the decision tree and the two ANNs as phenotype, could be represented by a simple genotype of two vectors, $w_1 \in \mathbb{R}^m$ and $w_2 \in \mathbb{R}^n$, showing the weights of the hidden layer in ANN-1 and ANN-2, respectively.

After making these application-specific design decisions, developing a suitable EA is rather straightforward because one can simply use the standard machinery and employ any EA that works in real-valued vector spaces. The only robotic-specific component is the fitness function that requires one or more test runs of the robot using the controller under evaluation. Then the behaviour of the robot is observed and some task-specific quality measure is taken to determine a fitness value. Running such an EA is not different from running any other EA, except for the fitness measurements, see Fig. 17.1. Sometimes, these can be performed by a simulator and the whole evolutionary process can be completed inside a computer. Alternatively, some or all fitness evaluations are carried out by real robots. In this case the genotype to be evaluated is sent to a robot to test it. Such a test delivers fitness values that are sent back to the computer running the EA. Employing real-world fitness evaluations implies a specific architecture, with a computational component that runs the EA and a robot component that executes the real-world fitness measurements. However, it could be argued that for the EA this dual architecture is invisible, it only needs to pause for longish periods waiting for the fitness values to be returned. A general approach to cope with time-consuming fitness evaluations is the use of surrogate models, as discussed in Chap. 14.

In terms of the problem categories discussed in Chap. 1, the task of designing a good robot controller represents a model-building problem. As explained above, a robot controller is a mapping between sensory inputs and actuator

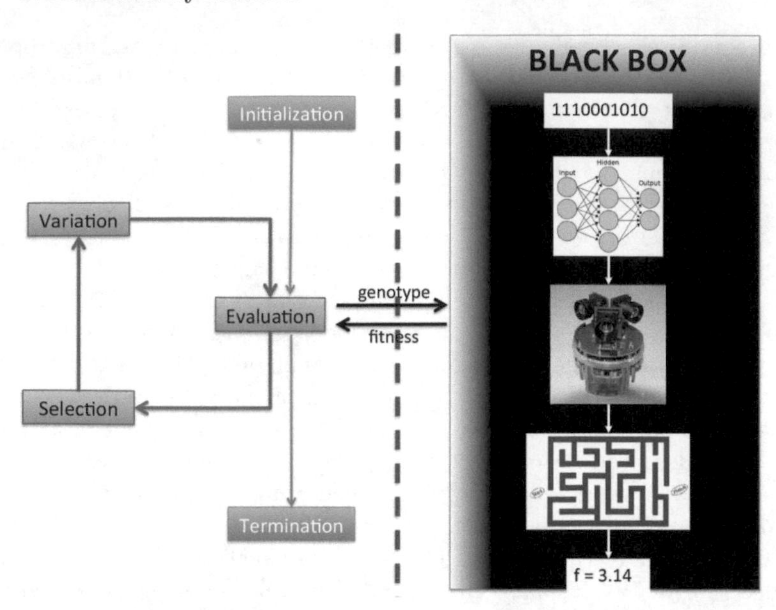

Fig. 17.1. Classic scheme of evolving robot designs. Note that any EA can be employed here, the only application-specific detail is the way fitness values are calculated

outputs and the essence of constructing a good controller is to find a mapping that provides an appropriate output for all inputs. In general, a model-building problem can be converted into an optimisation problem by creating a training set of input–output records and measuring model quality by the number or percentage of correct outputs it produces on the training set. For our example we could set up a training set consisting of pairs of sensory input patterns and corresponding actuator commands. However, this requires a microscopic-level description of desired robot behaviour and in practice this is not feasible. Instead, we could follow a macroscopic-level approach and use training cases where the input component describes the starting conditions (e.g., position of the robot in the arena and the brightness in the room) and the output component specifies that the robot is close enough to the light source.

17.3 Offline and Online Evolution of Robots

The usual approach in evolutionary robotics is to employ **offline evolution**, cf. Fig. 17.2. This means that an evolutionary algorithm to find a good controller is applied before the operational period of the robot. When the user is satisfied with the evolved controller, then it is deployed (installed on the physical robot) and the operational stage can start. In general, the evolved controllers do not change after deployment during the operational stage, or at

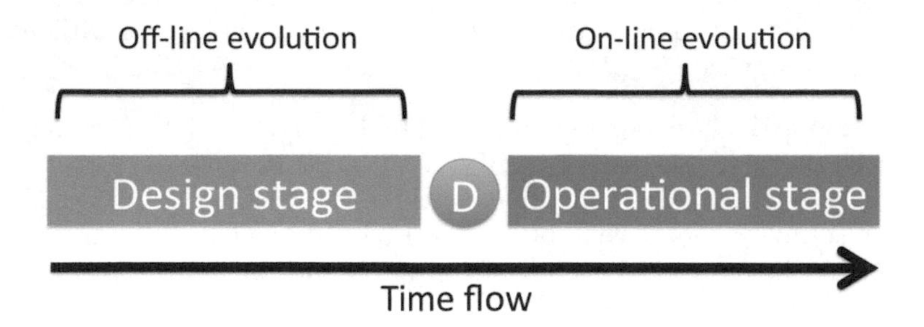

Fig. 17.2. Workflow of robot design distinguishing the design stage and the operational stage of the robots, separated by the moment of deployment (circle in the middle). Offline evolution takes place in the design stage and the evolved features (controllers or morphological details) do not change after deployment. Online evolution is performed during the operational period, which means that the robot's features are continually changed by the evolutionary operators

least not by evolutionary operators. The alternative is to apply **online evolution** to evolve controllers during the operational period. This implies that evolutionary operators can change the robot's control software even after its deployment.

The difference between the offline and online approaches was identified early in the history of the field, for instance by Walker et al. in [449] who used the names "training phase evolution" and "lifelong adaptation by evolution", respectively. However, the online variant has received very little attention and the huge majority of existing work is based on the offline approach.[2] This preference is understandable, because it is fully in line with the widespread use of EAs as optimizers. However, natural evolution is not a function optimizer. The natural role of evolution is that of permanent adaptation and this role is expected to become more and more important in the future of robotics, as phrased by Nelson et al. in [315].

> "Advanced autonomous robots may someday be required to negotiate environments and situations that their designers had not anticipated. The future designers of these robots may not have adequate expertise to provide appropriate control algorithms in the case that an unforeseen situation is encountered in a remote environment in which a robot

[2] This is true of machine learning in general – the huge majority of algorithms and papers about model building are concerned with learning from a fixed training set rather than continuous learning.

cannot be accessed. It is not always practical or even possible to define every aspect of an autonomous robots environment, or to give a tractable dynamical systems-level description of the task the robot is to perform. The robot must have the ability to learn control without human supervision."

The vision articulated here identifies a problem, and evolutionary computing can be (part of) the solution. Of course, online adaptation need not be based on or restricted to evolutionary techniques. One may even argue that evolution is too slow for such applications. However, we have seen in Chap. 11 that EAs can be used for dynamic optimization and there are good examples of using EAs for online learning, e.g., [75].

17.4 Evolutionary Robotics: The Problems Are Different

After the above considerations, we can discuss the properties that make robotics a very interesting context for evolutionary computing. To begin with, let us consider the problems, in particular the fitness functions for robotics. These exhibit a number of particular features that may not be unique one by one, but together they form a very challenging combination.

- The fitness function is very noisy.
- The fitness function is very costly.
- The fitness function is very complex.
- There may not be an explicit fitness function at all.
- The fitness landscape has 'no-go areas'.

The first four challenges occur in all segments of ER, including offline evolution based on simulation, the fifth one is characteristic for applications with online evolution is real hardware. In what follows we discuss these one by one.

Noisy Fitness: Noise is inherent in the physical world. For instance, two LEDs on a robot are not fully identical and the light they emit can be different even though they receive the same controller instruction. Furthermore, seemingly small environmental details can cause differences, e.g., the left wheel may be a bit warmer than the right wheel and make the robot deviate from the intended trajectory. This means that the link between actually controllable responses and robot behaviour is nondeterministic and the range of variations can be quite large.

Costly Fitness: Calculating the tour length for a given route of a traveling salesman can be done in the blink of an eye. However, measuring the time a robot needs for navigating to a target location can take several minutes. Moreover, it is an essential requirement for robot controllers that they work under many different circumstances. To this end, the controller must be tested under different initial conditions, for instance starting at different locations of the arena and/or under various lights. Ultimately this means that many time-consuming measurements are needed to assess the fitness of a given controller.

Complex Fitness: The phenotypes encoded by genotypes in ER are controllers that define the response of the actuators for any given combination of the inputs provided by the sensors.[3] However, a controller can only determine the actuator response on a low level (e.g., the torque on the left wheel or the colour of the LED light), whereas the fitness depends on exhibited robot behaviour on a high level (e.g., the robot is driving in circles). Hence, the phenotypes, i.e., the controllers, cannot be directly evaluated and one has to observe and assess the robot behavior induced by the given controller. Thus, in EC we have a 3-step chain, genotype–phenotype–fitness, while in ER the chain is 4-fold, genotype–phenotype–behavior–fitness. Furthermore, in the case of robot swarms, we have one more level, that of the group behaviour that raises further questions about considering the individual or the group when executing selection [448]. Thus, in general, the link between actually controllable values in the genotypes and robot behaviour is complex and ill-defined without analytical models.

Implicit Fitness: In the true spirit of biological evolution, EC can be used in an objective-free fashion to invent robots that are well suited to some (previously unknown and/or changing) environment. In such cases robots do not have a quantifiable level of fitness; they are fit if they survive long enough to be able to mate and propagate their genetic makeup. Such problems are very different from classic optimization and design problems in that they require a 'curiosity instinct' for pursuing novelty and open-ended exploration, rather than a drive for chasing optima.

'No-Go-Areas' in Fitness Landscape: Evaluating a poor candidate solution of a traveling salesman problem can waste time. Evaluating a poor candidate solution (bad controller) can destroy the robot. When working in real hardware, such candidate solutions must be avoided during the evolutionary search.

For further elaboration let us focus on a particular subarea in robotics: applications that involve (possibly large) groups of autonomous robots that undergo online evolution. One noteworthy property of such applications is that artificial evolution is required to play two roles at the same time: optimising towards some quantifiable skills of the robots as specified by the user as well as enabling open-ended adaptation to the given environment. To illustrate this recall the examples about Mars explorers and deep mining 'robot moles'. These are problems where the environment is not fully known to the designers in advance. The robot designs at the time of deployment are therefore just initial guesses and the designers' work needs to be finalised on the spot. As a necessary prerequisite for surviving at all, the robots must be fit for the environment (viability) and to fulfill the users' expectations they should

[3] Genotypes can of course also encode for morphological features. We disregard this option here for the sake of simplifying the argument.

perform their tasks well (usefulness). Artificial evolutionary systems today are typically developed with either but not both roles in mind. Optimising task performance is the leading motive in evolutionary computing and mainstream evolutionary robotics, whereas 'goal-less' adaptation to some environment is common in artificial life, but recent work has demonstrated that the two priorities can be integrated elegantly in one system [204].

Another special property here is that robots can be passive as well as active components of the evolutionary system.[4] On the one hand, robots are *passive* from the evolutionary algorithm perspective if they just passively undergo selection and reproduction. This is the dominant perspective when using traditional EAs to optimise robot designs in an offline fashion. In these cases the robots' only role in the EA is to establish fitness values that the selection mechanism needs in order to decide about who is to reproduce and who is to survive. On the other hand, robots can be *active* from the evolutionary algorithm perspective because they have processors on board that can perform computations and execute evolutionary operators. For instance, a robot could select another robot to mate with, perform recombination on the two parent controllers, evaluate the child controllers, and select the best one for further usage. Obviously, different robots can apply different reproduction operators and/or different selection preferences. This constitutes a new type of evolutionary system that is not only distributed, but also heterogeneous regarding the evolutionary operators.

A further property worth mentioning is that robots can also influence the evolutionary dynamics implicitly by structuring the population. This influence is grounded in the physical embedding which makes a group of robots spatially structured. Spatially structured EAs are, of course, nothing new, see for instance Sect. 5.5.7. However, in evolving robot swarms this structure is not designed, but emergent, without being explicitly specified by the EA designer, and it can have a large impact on the evolutionary process. For instance, the maximum sensor and communication ranges imply (dynamically changing) neighbourhoods that affect evolutionarily relevant interactions, e.g., mate selection and recombination. Obviously, sensor ranges and the like are robot attributes that can be controlled by the designer or experimenter, but their evolutionary effects are complex and to date there is no know-how available on adjusting evolutionary algorithms through such parameters.

To conclude this section let us briefly consider simulated and real fitness evaluations. A straightforward way of circumventing hardware-related difficulties and coping with costly fitness functions is the use of simulators. The simulator must represent the robot and the test environment including obstacles and other robots, if applicable. Using a simulator with a high level of abstraction (low level of details) can make a simulation orders of magnitude faster than a real-life experiment. Meanwhile, detailed simulations with a hi-fidelity physics engine and robots with many sensors can be much slower than

[4] This is not the same passive–active division we discussed in Sect. 17.1.

real life. Parallel execution on several processors can mitigate this problem and, in the end, simulations can save a lot of time when assessing the fitness of robot controllers. However, this comes at a price, the infamous **reality gap** that stands for the inevitable differences between simulated and real-life behaviour [233]. For evolutionary robotics, the reality gap means that the fitness function does not capture the real target correctly and even the high fitness solutions may perform poorly in the real world.

17.5 Evolutionary Robotics: The Algorithms Are Different

As mentioned above, standard EAs can be used to help design robots in an offline fashion. However, new types of EAs are needed when controllers, morphologies or both are evolved on the fly. Despite a handful of promising publications, there is not much know-how on such EAs, indicating the need for further research. In the following we try to identify some of the related issues, arranged by the following aspects:

- the overall system architecture (evolutionary vs. robotic components);
- the evolutionary operators (variation and selection);
- the fitness evaluations;
- the use of simulators.

By system architecture we mean the structural and functional relations between the evolutionary and the robotics system components as outlined in [123]. The classic offline architecture relies on a computer that hosts a population of genotypes that represent robot controllers, cf. Fig. 17.1. This computer executes the evolutionary operators (variation and selection), managed in the usual EC fashion. Fitness evaluations can be performed on this computer using a simulator. In this case the whole evolutionary process can be completed inside this computer. Alternatively, (some) fitness evaluations can be carried out by real robots. To this end, the genotype to be evaluated is sent to a robot to test it. After an evaluation period, the robot sends the fitness values back to the computer. Using more robots, evaluation of genotypes can happen in parallel which can lead to significant speed up. After finishing the evolutionary process the best genotype is deployed in the robot(s).

Online architectures are aimed at evolving robot controllers during their operational period. Here we can distinguish between centralised and distributed systems. One type of centralised system works as the online variant of the classic architecture. A master computer oversees the robots and runs the evolutionary process: it collects fitness information, executes variation and selection operators, and sends new controllers to the robots for them to use and test. As opposed to the offline scheme, the operational usage and fitness evaluation of controllers are not separated here. This is illustrated in Fig. 17.3, left. In the other type of centralised online system the computer running the evolutionary algorithm is internal to the robot, see Fig. 17.3, right.

Distributed online architectures also come in two flavours. In a pure distributed system each robot carries one genome that encodes its own phenotypic features. Selection and reproduction require interaction between more robots, see Fig. 17.4, left. In a hybrid system the encapsulated and the pure distributed approaches are combined, as shown in Fig. 17.4, right.

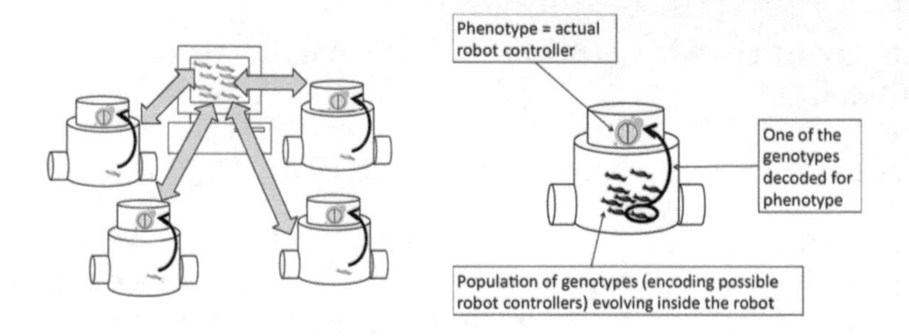

Fig. 17.3. Centralised online evolution architectures. Left: an external master computer oversees the robots and runs the evolutionary process. Right: an internal computer (the robot's own processor) runs an encapsulated evolutionary process, where selection and reproduction do not require information from other robots

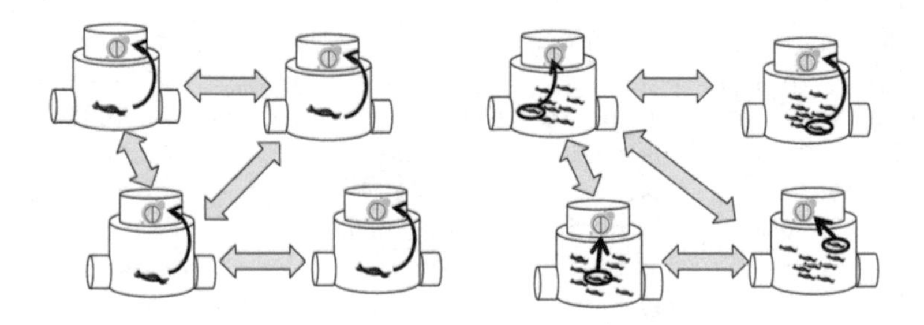

Fig. 17.4. Distributed online evolution architectures. Left: distributed system with one-robot–one-genome. Selection and reproduction require interaction between more robots. Right: a hybrid system is encapsulated and distributed. Encapsulated populations form islands that evolve independently but can crossfertilize.

Regarding the evolutionary operators, we can mention two prominent requirements. First and foremost, reproduction must be constrained to physically viable candidate solutions. This requirement is related to the issue of the 'no-go areas' on the fitness landscape: a poor candidate solution (a bad controller) can break, immobilize, or otherwise disable the robot. Hence, the evolutionary path should avoid these regions. The area of evolutionary constraint handling, cf. Chap. 13, could be relevant here, especially the use of intelligent reproduction operators and repair mechanisms. However, these are always based on application-specific heuristics and as of today there are no general recipes in EC that would offer solutions here. One promising avenue to avoid damaging robots is the use of a surrogate model to estimate the viability of new candidate solutions before launching them in the real world. We discussed the use of surrogates for interactive EAs in Sect. 14.2. Such models can be maintained and continually updated in every robot independently, but the robot population could also collaborate and develop the models collectively. The second requirement concerns the speed of evolution. Because of the online character of the robotic application, rapid progress is desirable. In other words, it should not take many iterations to reach good performance levels. To this end, there are generic options supported by common evolutionary computing wisdom. For example, using high selection pressure or multiparent crossovers can be considered as accelerators, but with a caveat: greedy search can easily get trapped in local optima.

There are also special considerations for selection in the distributed cases. The physical embedding implies a natural notion of distance, that in turn induces neighbourhoods for robots. The composition, location, and size of these neighbourhoods will depend on the specific circumstances, e.g., ranges for sensing and communication, the environment, noise, etc., but in general it will hold that robots do not have direct access to all other robots. Thus, the selection operators will act on local and partial information, using fitness data of only a fraction of the whole population. In principle, this can be circumvented by using some form of epidemic or gossiping protocols that help in estimating global information. The first results with autonomous selection in peer-to-peer evolutionary systems are very promising [463].

Concerning fitness evaluations, perhaps the most important robotics-specific feature is their duration. In typical EAs for optimization or for offline robot design the real-time duration of fitness evaluations does not impact the evolutionary dynamics. Figuratively speaking, the EA can freeze until the fitness values are returned and proceed with the search afterwards. Thus, no matter how long a fitness evaluation really takes, from the EA perspective it can be seen as instantaneous. Using online evolution in robots this is different, because the real world will not freeze awaiting the fitness evaluations to finish. In such systems, the time needed to complete fitness evaluations is a parameter with a paramount impact. On the one hand, long evaluations imply fewer evaluations in any given time interval. This is certainly disadvantageous for the (real-time) speed of the evolutionary progress. On the other hand, short evaluations increase the risk of poor fitness estimations. This misleads selection operators and drives the population to suboptimal regions. Consequently, the parameter(s) that determine the duration of fitness evaluations should have robust parameter values that work over a wide range of circumstances, or a good parameter control method should be applied that adjusts the values appropriately on the fly. Research in this direction is just starting, but promising results have been published recently based on a (self-)adaptive mechanism [117].

Finally, let us return to the issue of using simulations to assess the quality of robot designs. A very promising idea here is to use simulations *within* robots for continuous self-modelling [60]. In principle, this means that fitness evaluations can be performed without real-life trials during an online evolutionary process. This approach bears great advantages because it can save time, it can save robots (that otherwise may run into irreparable problems), and it can help experimenters learn about the problem (if only these models are readable). A very recent study showing the benefits of this approach in a robot swarm undergoing online evolution is presented in [323].

17.6 A Glimpse into the Future

Let us conclude this chapter with a somewhat speculative section about the future of evolutionary robotics and evolutionary computing itself. We believe that the combination of robotics and evolutionary computing has a great potential with mutual benefits. On the one hand, EC can provide solutions to hard problems in robotics. In particular, evolution can help design and optimise robot controllers, robot morphologies, or both and it can provide adaptive capabilities for on-the-fly adjustments without human intervention. On the other hand, robotics forms a demanding testing ground for EC and the specific (combination of) challenges in robotics can drive the development of novel types of artificial evolutionary systems.

A detailed discussion of the robotics perspective is beyond the scope of this book. For the interested reader, we recommend [119], which distinguishes four major areas of development at different levels of maturity:

- automatic parameter tuning in the design space (mature technique),
- evolutionary-aided design (current trend),
- online evolutionary adaptation (current trend),
- automatic synthesis (long-term research).

Regarding EC, our main premise is that we are at the verge of a major breakthrough that will open up a completely new substrate for artificial evolutionary systems. This breakthrough is the technology of self-reproducing physical artefacts. This may be achieved through, for instance, 3D printing or autonomous self-assembly in the very near future. This achievement would extend the meaning of evolvable hardware [466]. To be specific, this means a technology that enables the autonomous construction of a new robot based on a genetic plan produced by mutation or crossover applied to the genetic plan(s) of its parent(s). Given this technology it will be possible to make systems where bodies and brains (i.e., morphologies and controllers) of robots coevolve without the human in the loop. The *Triangle of Life* framework of

Eiben et al. [132] provides a general description of such systems and the paper demonstrates a rudimentary implementation of some of the main system components with cubic robots as building blocks. The Triangle of Life can be considered as the counterpart of the general scheme of an evolutionary algorithm in Chap. 3.

From a historical perspective, the technology of self-reproducing physical artefacts will be a radical game changer in that it will allow us to create, utilize, and study artificial evolutionary processes outside of computers. The entire history of EC can be considered as the process of learning how to construct and operate artificial evolutionary systems in imaginary, digital spaces. This very book is a collection of knowledge accumulated over the last few decades about this. The technology of self-reproducing artifacts will bring the game into real physical spaces. The emerging area will be concerned with Embodied Artificial Evolution [136] or the Evolution of Things [122]. The historical perspective is illustrated by Fig. 17.5 that shows two major transitions of Darwinian principles from one substrate to another. The first transition in the 20th century followed the emergence of computer technology that provided the possibility of creating digital worlds that are very flexible and controllable. This brought about the opportunity to become active masters of evolutionary processes that are fully designed and executed by human experimenters. The second transition is about to take place through an emerging technology based on material science, rapid prototyping (3D printing) and evolvable hardware. This will provide flexible and controllable physical, *in materio*, substrates and be the basis of Embodied Artificial Evolution or the Evolution of Things [147].

Finally, let us mention a new kind of synergy between evolutionary computing, artificial life, robotics, and biology. As of today, the level of technology will not allow for emulating all chemical and biological micro-mechanisms underlying evolution in some artificial substrate. However, even a system that only mimics the macro-mechanisms (e.g., selection, reproduction, and heredity) in a physical medium is a better tool for studying evolution than pure software simulations because it will not violate the laws of physics and will be able to exploit the richness of matter. A recently published list of grand challenges for evolutionary robotics includes "Open-Ended Robot Evolution" where physical robots undergo open-ended evolution in an open environment [121]. With such systems one could investigate fundamental issues, e.g., the minimal conditions for evolution to take place, the factors influencing evolvability, or the rate of progress under various circumstances. Given enough time, we may even witness some of the events natural evolution encountered, such as the emergence of species, perhaps even the Cambrian explosion. This line of development can bridge evolutionary computing, artificial life, robotics, and biology, producing a new category: Life, but not as we know it.

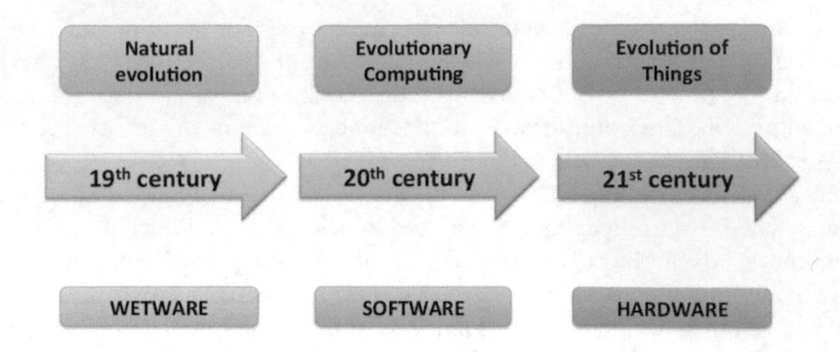

Fig. 17.5. Major transitions of evolutionary principles from one substrate to another. Natural evolution *in vivo* was discovered and described in the 19th century. Computer technology in the 20th century provided digital, *in silico* substrates for creating, utilizing, and studying artificial evolutionary processes. This formed the basis of Evolutionary Computing. An emerging technology based on material science, rapid prototyping (3D printing) and evolvable hardware will provide physical, *in materio* substrates. This will be the basis of Embodied Artificial Evolution or the Evolution of Things.

For exercises and recommended reading for this chapter, please visit `www.evolutionarycomputation.org`.

References

1. E.H.L. Aarts, A.E. Eiben, and K.M. van Hee. A general theory of genetic agorithms. Technical Report 89/08, Eindhoven University of Technology, 1989.
2. E.H.L. Aarts and J. Korst. *Simulated Annealing and Boltzmann Machines.* Wiley, Chichester, UK, 1989.
3. E.H.L. Aarts and J.K. Lenstra, editors. *Local Search in Combinatorial Optimization.* Discrete Mathematics and Optimization. Wiley, Chichester, UK, June 1997.
4. D. Adams. *The Hitchhiker's Guide to the Galaxy.* Guild Publishing, London, 1986.
5. E. Alba and B. Dorronsoro. *Cellular Genetic Algorithms.* Computational Intelligence and Complexity. Springer, 2008.
6. L. Altenberg. The schema theorem and Price's theorem. In Whitley and Vose [462], pages 23–50.
7. D. Andre and J.R. Koza. Parallel genetic programming: A scalable implementation using the transputer network architecture. In P.J. Angeline and K.E. Kinnear, editors, *Advances in Genetic Programming 2*, pages 317–338. MIT Press, Cambridge, MA, 1996.
8. P.J. Angeline. Adaptive and self-adaptive evolutionary computations. In *Computational Intelligence*, pages 152–161. IEEE Press, 1995.
9. P.J. Angeline. Subtree crossover: Building block engine or macromutation? In Koza et al. [256], pages 9–17.
10. P.J. Angeline. Competitive fitness evaluation. In Bäck et al. [28], chapter 3, pages 12–14.
11. J. Arabas, Z. Michalewicz, and J. Mulawka. GAVaPS – a genetic algorithm with varying population size. In ICEC-94 [229], pages 73–78.
12. D. Ashlock. *Evolutionary Computation for Modeling and Optimization.* Springer, 2006.
13. Anne Auger, Steffen Finck, Nikolaus Hansen, and Raymond Ros. BBOB 2010: Comparison Tables of All Algorithms on All Noiseless Functions. Technical Report RT-388, INRIA, September 2010.
14. R. Axelrod. *The Evolution of Cooperation.* Basic Books, New York, 1984.
15. R. Axelrod. The evolution of strategies in the iterated prisoner's dilemma. In L. Davis, editor, *Genetic Algorithms and Simulated Annealing.* Pitman, London, 1987.

16. J. Bacardit, M. Stout, J.D. Hirst, K. Sastry, X. Llor, and N. Krasnogor. Automated alphabet reduction method with evolutionary algorithms for protein structure prediction. In Bosman et al. [65], pages 346–353.
17. T. Bäck. The interaction of mutation rate, selection and self-adaptation within a genetic algorithm. In Männer and Manderick [282], pages 85–94.
18. T. Bäck. Self adaptation in genetic algorithms. In Varela and Bourgine [440], pages 263–271.
19. T. Bäck. Selective pressure in evolutionary algorithms: A characterization of selection mechanisms. In ICEC-94 [229], pages 57–62.
20. T. Bäck. Generalised convergence models for tournament and (μ, λ) selection. In Eshelman [156], pages 2–8.
21. T. Bäck. Order statistics for convergence velocity analysis of simplified evolutionary algorithms. In Whitley and Vose [462], pages 91–102.
22. T. Bäck. *Evolutionary Algorithms in Theory and Practice.* Oxford University Press, Oxford, UK, 1996.
23. T. Bäck, editor. *Proceedings of the 7th International Conference on Genetic Algorithms.* Morgan Kaufmann, San Francisco, 1997.
24. T. Bäck. Self-adaptation. In Bäck et al. [28], chapter 21, pages 188–211.
25. T. Bäck, A.E. Eiben, and N.A.L. van der Vaart. An empirical study on GAs "without parameters". In Schoenauer et al. [368], pages 315–324.
26. T. Bäck, D.B. Fogel, and Z. Michalewicz, editors. *Handbook of Evolutionary Computation.* Institute of Physics Publishing, Bristol, and Oxford University Press, New York, 1997.
27. T. Bäck, D.B. Fogel, and Z. Michalewicz, editors. *Evolutionary Computation 1: Basic Algorithms and Operators.* Institute of Physics Publishing, Bristol, 2000.
28. T. Bäck, D.B. Fogel, and Z. Michalewicz, editors. *Evolutionary Computation 2: Advanced Algorithms and Operators.* Institute of Physics Publishing, Bristol, 2000.
29. T. Bäck and Z. Michalewicz. Test landscapes. In Bäck et al. [26], chapter B2.7, pages 14–20.
30. T. Bäck and H.-P. Schwefel. An overview of evolutionary algorithms for parameter optimization. *Evolutionary Computation*, 1(1):1–23, 1993.
31. T. Bäck, D. Vermeulen, and A.E. Eiben. Effects of tax and evolution in an artificial society. In H.J. Caulfield, S.H. Chen, H.D. Cheng, R. Duro, V. Honavar, E.E. Kerre, M. Lu, M.G. Romay, T.K. Shih, D. Ventura, P. Wang, and Y. Yang, editors, *Proceedings of the Sixth Joint Conference on Information Sciences, (JCIS 2002)*, pages 1151–1156. JCIS/Association for Intelligent Machinery, 2002.
32. J.E. Baker. Reducing bias and inefficiency in the selection algorithm. In Grefenstette [198], pages 14–21.
33. P. Balaprakash, M. Birattari, and T. Stützle. Improvement strategies for the F-Race algorithm: Sampling design and iterative refinementace algorithm: Sampling design and iterative refinement. In T. Bartz-Beielstein, M. Blesa Aguilera, C. Blum, B. Naujoks, A. Roli, G. Rudolph, and M. Sampels, editors, *Hybrid Metaheuristics*, volume 4771 of *Lecture Notes in Computer Science*, pages 108–122. Springer, 2007.
34. J.M. Baldwin. A new factor in evolution. *American Naturalist*, 30, 1896.

35. Shummet Baluja. Population-based incremental learning: A method for integrating genetic search based function optimization and competitive learning. Technical report, Carnegie Mellon University, Pittsburgh, PA, USA, 1994.
36. W. Banzhaf, J. Daida, A.E. Eiben, M.H. Garzon, V. Honavar, M. Jakiela, and R.E. Smith, editors. *Proceedings of the Genetic and Evolutionary Computation Conference (GECCO-1999).* Morgan Kaufmann, San Francisco, 1999.
37. W. Banzhaf, P. Nordin, R.E. Keller, and F.D. Francone. *Genetic Programming: An Introduction.* Morgan Kaufmann, San Francisco, 1998.
38. W. Banzhaf and C. Reeves, editors. *Foundations of Genetic Algorithms 5.* Morgan Kaufmann, San Francisco, 1999.
39. N.A. Baricelli. Numerical testing of evolution theories, part 1. *Acta Biotheor.*, 16:69–98, 1962.
40. Abu S. S. M. Barkat Ullah, Ruhul Sarker, David Cornforth, and Chris Lokan. AMA: a new approach for solving constrained real-valued optimization problems. *Soft Computing*, 13(8-9):741–762, August 2009.
41. T. Bartz-Beielstein. *New Experimentalism Applied to Evolutionary Computation.* PhD thesis, Universität Dortmund, 2005.
42. T. Bartz-Beielstein, K.E. Parsopoulos, and M.N. Vrahatis. Analysis of Particle Swarm Optimization Using Computational Statistics. In Chalkis, editor, *Proceedings of the International Conference of Numerical Analysis and Applied Mathematics (ICNAAM 2004)*, pages 34–37. Wiley, 2004.
43. T. Bartz-Beielstein and M. Preuss. Considerations of Budget Allocation for Sequential Parameter Optimization (SPO). In L. Paquete et al., editors, *Workshop on Empirical Methods for the Analysis of Algorithms, Proceedings*, pages 35–40, Reykjavik, Iceland, 2006. Online Proceedings.
44. J.C. Bean. Genetic algorithms and random keys for sequencing and optimisation. *ORSA Journal of Computing*, 6(2):154–160, 1994.
45. J.C. Bean and A.B. Hadj-Alouane. A dual genetic algorithm for bounded integer problems. Technical Report 92-53, University of Michigan, 1992.
46. R.K. Belew and L.B. Booker, editors. *Proceedings of the 4th International Conference on Genetic Algorithms.* Morgan Kaufmann, San Francisco, 1991.
47. P. Bentley. From coffee tables to hospitals: Generic evolutionary design. In Bentley [48], pages 405–423.
48. P.J. Bentley, editor. *Evolutionary Design by Computers.* Morgan Kaufmann, San Francisco, 1999.
49. P.J. Bentley and D.W. Corne, editors. *Creative Evolutionary Systems.* Morgan Kaufmann, San Francisco, 2002.
50. P.J. Bentley and D.W. Corne. An introduction to creative evolutionary systems. In Bentley and Corne [49], pages 1–75.
51. A.D. Bethke. *Genetic Algorithms as Function Optimizers.* PhD thesis, University of Michigan, 1981.
52. H.-G. Beyer. *The Theory of Evolution Strategies.* Springer, 2001.
53. H.-G. Beyer and D.V. Arnold. Theory of evolution strategies: A tutorial. In Kallel et al. [240], pages 109–134.
54. H.-G. Beyer and H.-P. Schwefel. Evolution strategies: A comprehensive introduction. *Natural Computing*, 1(1):3–52, 2002.
55. M. Birattari. *Tuning Metaheuristics.* Springer, 2005.
56. M. Birattari, Z. Yuan, P. Balaprakash, and T. Stützle. F-Race and iterated F-Race: An overview. In T. Bartz-Beielstein, M. Chiarandini, L. Paquete, and

M. Preuss, editors, *Experimental Methods for the Analysis of Optimization Algorithms*, pages 311–336. Springer, 2010.

57. S. Blackmore. *The Meme Machine.* Oxford University Press, Oxford, UK, 1999.
58. T. Blickle and L. Thiele. A comparison of selection schemes used in genetic algorithms. Technical Report TIK Report 11, December 1995, Computer Engineering and Communication Networks Lab, Swiss Federal Institute of Technology, 1995.
59. J.S. De Bonet, C. Isbell, and P. Viola. Mimic: Finding optima by estimating probability densities. *Advances in Neural Information Processing Systems*, 9:424–431, 1997.
60. J. Bongard, V. Zykov, and H. Lipson. Resilient machines through continuous self-modeling. *Science*, 314:1118–1121, 2006.
61. J.C. Bongard. Evolutionary robotics. *Communications of the ACM*, 56(8), 2013.
62. L.B. Booker. *Intelligent Behaviour as an adaptation to the task environment.* PhD thesis, University of Michigan, 1982.
63. Y. Borenstein and A. Moraglio, editors. *Theory and Principled Methods for the Design of Metaheuristics.* Natural Computing Series. Springer, 2014.
64. P. Bosman and D. Thierens. Expanding from discrete to continuous estimation of distribution algorithms: The idea. In Schoenauer et al. [368], pages 767–776.
65. Peter A. N. Bosman, Tina Yu, and Anikó Ekárt, editors. *GECCO '07: Proceedings of the 2007 GECCO conference on Genetic and evolutionary computation*, New York, NY, USA, 2007. ACM.
66. M.F. Bramlette. Initialization, mutation and selection methods in genetic algorithms for function optimization. In Belew and Booker [46], pages 100–107.
67. H.J. Bremermann, M. Rogson, and S. Salaff. Global properties of evolution processes. In H.H. Pattee, E.A. Edlsack, L. Fein, and A.B. Callahan, editors, *Natural Automata and Useful Simulations*, pages 3–41. Spartan Books, Washington DC, 1966.
68. L. Bull. *Artificial Symbiology.* PhD thesis, University of the West of England, 1995.
69. L. Bull and T.C. Fogarty. Horizontal gene transfer in endosymbiosis. In C.G. Langton and K. Shimohara, editors, *Proceedings of the 5th International Workshop on Artificial Life: Synthesis and Simulation of Living Systems (ALIFE-96)*, pages 77–84. MIT Press, Cambridge, MA, 1997.
70. L. Bull, O. Holland, and S. Blackmore. On meme–gene coevolution. *Artificial Life*, 6:227–235, 2000.
71. E. K. Burke, T. Curtois, M. R. Hyde, G. Kendall, G. Ochoa, S. Petrovic, J.A. Vázquez Rodríguez, and M. Gendreau. Iterated local search vs. hyper-heuristics: Towards general-purpose search algorithms. In *IEEE Congress on Evolutionary Computation*, pages 1–8. IEEE Press, 2010.
72. E.K. Burke, G. Kendall, and E. Soubeiga. A tabu search hyperheuristic for timetabling and rostering. *Journal of Heuristics*, 9(6), 2003.
73. E.K. Burke and J.P. Newall. A multi-stage evolutionary algorithm for the timetable problem. *IEEE Transactions on Evolutionary Computation*, 3(1):63–74, 1999.
74. E.K. Burke, J.P. Newall, and R.F. Weare. Initialization strategies and diversity in evolutionary timetabling. *Evolutionary Computation*, 6(1):81–103, 1998.

75. M.V. Butz. *Rule-Based Evolutionary Online Learning Systems*. Studies in Fuzziness and Soft Computing Series. Springer, 2006.
76. P. Caleb-Solly and J.E. Smith. Adaptive surface inspection via interactive evolution. *Image and Vision Computing*, 25(7):1058–1072, 2007.
77. A. Caponio, G.l. Cascella, F. Neri., N. Salvatore., and M. Sumner. A Fast Adaptive Memetic Algorithm for Online and Offline Control Design of PMSM Drives. *IEEE Transactions on Systems, Man, and Cybernetics, Part B: Cybernetics*, 37(1):28–41, 2007.
78. U.K. Chakraborty. An analysis of selection in generational and steady state genetic algorithms. In *Proceedings of the National Conference on Molecular Electronics*. NERIST (A.P.) India, 1995.
79. U.K. Chakraborty, K. Deb, and M. Chakraborty. Analysis of selection algorithms: A Markov Chain aproach. *Evolutionary Computation*, 4(2):133–167, 1997.
80. P. Cheeseman, B. Kenefsky, and W. M. Taylor. Where the really hard problems are. In *Proceedings of the Twelfth International Joint Conference on Artificial Intelligence, IJCAI-91*, pages 331–337, 1991.
81. K. Chellapilla and D.B. Fogel. Evolving an expert checkers playing program without human expertise. *IEEE Transactions on Evolutionary Computation*, 5(4):422–428, 2001.
82. X. Chen. An Algorithm Development Environment for Problem-Solving. In *(ICCP), 2010 International Conference on Computational Problem-Solving*, pages 85–90, 2010.
83. Y.P. Chen. *Extending the Scalability of Linkage Learning Genetic Algorithms: - Theory & Practice*, volume 190 of *Studies in Fuzziness and Soft Computing*. Springer, 2006.
84. H. Cobb. An investigation into the use of hypermutation as an adaptive operator in a genetic algorithm having continuous, time-dependent nonstationary environments. Memorandum 6760, Naval Research Laboratory, 1990.
85. H.G. Cobb and J.J. Grefenstette. Genetic algorithms for tracking changing environments. In Forrest [176], pages 523–530.
86. C.A. Coello Coello, D.A. Van Veldhuizen, and G.B. Lamont. *Evolutionary Algorithms for Solving Multi-Objective Problems*. Kluwer Academic Publishers, Boston, 2nd edition, 2007. ISBN 0-3064-6762-3.
87. J.P. Cohoon, S.U. Hedge, W.N. Martin, and D. Richards. Punctuated equilibria: A parallel genetic algorithm. In Grefenstette [198], pages 148–154.
88. J.P. Cohoon, W.N. Martin, and D.S. Richards. Genetic algorithms and punctuated equilibria in VLSI. In Schwefel and Männer [374], pages 134–144.
89. P. Cowling, G. Kendall, and E. Soubeiga. A hyperheuristic approach to scheduling a sales summit. *Lecture Notes in Computer Science*, 2079:176–95, 2001.
90. B. Craenen, A.E. Eiben, and J.I. van Hemert. Comparing evolutionary algorithms on binary constraint satisfaction problems. *IEEE Transactions on Evolutionary Computation*, 7(5):424–444, 2003.
91. M. Crepinsek, S. Liu, and M. Mernik. Exploration and exploitation in evolutionary algorithms: A survey. *ACM Computing Surveys*, 45(3):35:1–35:33, July 2013.
92. C. Darwin. *The Origin of Species*. John Murray, 1859.

93. R. Das and D. Whitley. The only challenging problems are deceptive: Global search by solving order-1 hyperplanes. In Belew and Booker [46], pages 166–173.
94. D. Dasgupta and D. McGregor. SGA: A structured genetic algorithm. Technical Report IKBS-8-92, University of Strathclyde, 1992.
95. Y. Davidor. A naturally occurring niche & species phenomenon: The model and first results. In Belew and Booker [46], pages 257–263.
96. Y. Davidor, H.-P. Schwefel, and R. Männer, editors. *Proceedings of the 3rd Conference on Parallel Problem Solving from Nature*, number 866 in Lecture Notes in Computer Science. Springer, 1994.
97. L. Davis. Adapting operator probabilities in genetic algorithms. In Schaffer [365], pages 61–69.
98. L. Davis, editor. *Handbook of Genetic Algorithms.* Van Nostrand Reinhold, 1991.
99. T.E. Davis and J.C. Principe. A Markov chain framework for the simple genetic algorithm. *Evolutionary Computation*, 1(3):269–288, 1993.
100. R. Dawkins. *The Selfish Gene.* Oxford University Press, Oxford, UK, 1976.
101. R. Dawkins. *The Blind Watchmaker.* Longman Scientific and Technical, 1986.
102. K.A. De Jong. *An Analysis of the Behaviour of a Class of Genetic Adaptive Systems.* PhD thesis, University of Michigan, 1975.
103. K.A. De Jong. Genetic algorithms are NOT function optimizers. In Whitley [457], pages 5–18.
104. K.A. De Jong. *Evolutionary Computation: A Unified Approach.* The MIT Press, 2006.
105. K.A. De Jong and J. Sarma. Generation gaps revisited. In Whitley [457], pages 19–28.
106. K.A. De Jong and J. Sarma. On decentralizing selection algoritms. In Eshelman [156], pages 17–23.
107. K.A. De Jong and W.M. Spears. An analysis of the interacting roles of population size and crossover in genetic algorithms. In Schwefel and Männer [374], pages 38–47.
108. K.A. De Jong and W.M. Spears. A formal analysis of the role of multi-point crossover in genetic algorithms. *Annals of Mathematics and Artificial Intelligence*, 5(1):1–26, April 1992.
109. K. Deb. Genetic algorithms in multimodal function optimization. Master's thesis, University of Alabama, 1989.
110. K. Deb. *Multi-objective Optimization using Evolutionary Algorithms.* Wiley, Chichester, UK, 2001.
111. K. Deb and R.B. Agrawal. Simulated binary crossover for continuous search space. *Complex Systems*, 9:115–148, 1995.
112. K. Deb, S. Agrawal, A. Pratab, and T. Meyarivan. A Fast Elitist Non-Dominated Sorting Genetic Algorithm for Multi-Objective Optimization: NSGA-II. In Schoenauer et al. [368], pages 849–858.
113. K. Deb and H.-G. Beyer. Self-Adaptive Genetic Algorithms with Simulated Binary Crossover. *Evolutionary Computation*, 9(2), June 2001.
114. K. Deb and D.E. Goldberg. An investigation of niche and species formation in genetic function optimization. In Schaffer [365], pages 42–50.
115. E.D. deJong, R.A. Watson, and J.B. Pollack. Reducing bloat and promoting diversity using multi-objective methods. In Spector et al. [415], pages 11–18.

116. D. Dennett. *Darwin's Dangerous Idea*. Penguin,London, 1995.
117. C.M. Dinu, P. Dimitrov, B. Weel, and A. E. Eiben. Self-adapting fitness evaluation times for on-line evolution of simulated robots. In *GECCO '13: Proc of the 15th conference on Genetic and Evolutionary Computation*, pages 191–198. ACM Press, 2013.
118. B. Doerr, E. Happ, and C. Klein. Crossover can provably be useful in evolutionary computation. *Theor. Comput. Sci.*, 425:17–33, March 2012.
119. S. Doncieux, J.-B. Mouret, N. Bredeche, and V. Padois. Evolutionary robotics: Exploring new horizons. In S. Doncieux, N. Bredeche, and J.-B. Mouret, editors, *New Horizons in Evolutionary Robotics*, volume 341 of *Studies in Computational Intelligence*, chapter 2, pages 3–25. Springer, 2011.
120. S. Droste, T. Jansen, and I. Wegener. Upper and lower bounds for randomized search heuristics in black-box optimization. *Theory of Computing Systems*, 39(4):525–544, 2006.
121. A. E. Eiben. Grand challenges for evolutionary robotics. *Frontiers in Robotics and AI*, 1(4), 2014.
122. A. E. Eiben. In Vivo Veritas: towards the Evolution of Things. In T. Bartz-Beielstein, J. Branke, B. Filipič, and J. Smith, editors, *Parallel Problem Solving from Nature - PPSN XIII*, volume 8672 of *LNCS*, pages 24–39. Springer, 2014.
123. A. E. Eiben, E. Haasdijk, and N. Bredeche. Embodied, on-line, on-board evolution for autonomous robotics. In P. Levi and S. Kernbach, editors, *Symbiotic Multi-Robot Organisms: Reliability, Adaptability, Evolution*, chapter 5.2, pages 361–382. Springer, 2010.
124. A. E. Eiben and M. Jelasity. A critical note on Experimental Research Methodology in Experimental research methodology in EC. In *Proceedings of the 2002 Congress on Evolutionary Computation (CEC 2002)*, pages 582–587. IEEE Press, Piscataway, NJ, 2002.
125. A. E. Eiben and S. K. Smit. Evolutionary algorithm parameters and methods to tune them. In Y. Hamadi, E. Monfroy, and F. Saubion, editors, *Autonomous Search*, pages 15–36. Springer, 2012.
126. A.E. Eiben. Multiparent recombination. In Bäck et al. [27], chapter 33.7, pages 289–307.
127. A.E. Eiben. Evolutionary algorithms and constraint satisfaction: Definitions, survey, methodology, and research directions. In Kallel et al. [240], pages 13–58.
128. A.E. Eiben. Multiparent recombination in evolutionary computing. In Ghosh and Tsutsui [184], pages 175–192.
129. A.E. Eiben, E.H.L. Aarts, and K.M. Van Hee. Global convergence of genetic algorithms: a Markov chain analysis. In Schwefel and Männer [374], pages 4–12.
130. A.E. Eiben and T. Bäck. An empirical investigation of multi-parent recombination operators in evolution strategies. *Evolutionary Computation*, 5(3):347–365, 1997.
131. A.E. Eiben, T. Bäck, M. Schoenauer, and H.-P. Schwefel, editors. *Proceedings of the 5th Conference on Parallel Problem Solving from Nature*, number 1498 in Lecture Notes in Computer Science. Springer, 1998.
132. A.E. Eiben, N. Bredeche, M. Hoogendoorn, J. Stradner, J. Timmis, A.M. Tyrrell, and A. Winfield. The triangle of life: Evolving robots in real-time and real-space. In P. Lio, O. Miglino, G. Nicosia, S. Nolfi, and M. Pavone, editors, *Proc. 12th European Conference on the Synthesis and Simulation of Living Systems*, pages 1056–1063. MIT Press, 2013.

133. A.E. Eiben, R. Hinterding, and Z. Michalewicz. Parameter Control in Evolutionary Algorithms. *IEEE Transactions on Evolutionary Computation*, 3(2):124–141, 1999.
134. A.E. Eiben, B. Jansen, Z. Michalewicz, and B. Paechter. Solving CSPs using self-adaptive constraint weights: how to prevent EAs from cheating. In Whitley et al. [453], pages 128–134.
135. A.E. Eiben and M. Jelasity. A Critical Note on Experimental Research Methodology in EC. In *Proceedings of the 2002 IEEE Congress on Evolutionary Computation (CEC 2002)*, pages 582–587. IEEE Press, 2002.
136. A.E. Eiben, S. Kernbach, and E. Haasdijk. Embodied artificial evolution – artificial evolutionary systems in the 21st century. *Evolutionary Intelligence*, 5(4):261–272, 2012.
137. A.E. Eiben and Z. Michalewicz, editors. *Evolutionary Computation*. IOS Press, 1998.
138. A.E. Eiben, R. Nabuurs, and I. Booij. The Escher evolver: Evolution to the people. In Bentley and Corne [49], pages 425–439.
139. A.E. Eiben, P.-E. Raué, and Zs. Ruttkay. Repairing, adding constraints and learning as a means of improving GA performance on CSPs. In J.C. Bioch and S.H. Nienhuiys-Cheng, editors, *Proceedings of the 4th Belgian-Dutch Conference on Machine Learning*, number 94-05 in EUR-CS, pages 112–123. Erasmus University Press, 1994.
140. A.E. Eiben and G. Rudolph. Theory of evolutionary algorithms: a bird's eye view. *Theoretical Computer Science*, 229(1–2):3–9, 1999.
141. A.E. Eiben and Zs. Ruttkay. Constraint-satisfaction problems. In T. Baeck, D.B. Fogel, and Z. Michalewicz, editors, *Evolutionary Computation 2: Advanced Algorithms and Operators*, pages 75–86. Institute of Physics Publishing, 2000.
142. A.E. Eiben and A. Schippers. On evolutionary exploration and exploitation. *Fundamenta Informaticae*, 35(1-4):35–50, 1998.
143. A.E. Eiben and C.A. Schippers. Multi-parent's niche: n-ary crossovers on NK-landscapes. In Voigt et al. [445], pages 319–328.
144. A.E. Eiben, M.C. Schut, and A.R. de Wilde. Is self-adaptation of selection pressure and population size possible? A case study. In Thomas Philip Runarsson, Hans-Georg Beyer, Edmund K. Burke, Juan J. Merelo Guervós, L. Darrell Whitley, and Xin Yao, editors, *PPSN*, volume 4193 of *Lecture Notes in Computer Science*, pages 900–909. Springer, 2006.
145. A.E. Eiben and S. K. Smit. Parameter tuning for configuring and analyzing evolutionary algorithms. *Swarm and Evolutionary Computation*, 1(1):19–31, 2011.
146. A.E. Eiben and J.E. Smith. *Introduction to Evolutionary Computing*. Springer, Berlin Heidelberg, 2003.
147. A.E. Eiben and J.E. Smith. From evolutionary computation to the evolution of things. *Nature*, 2015. In press.
148. A.E. Eiben, I.G. Sprinkhuizen-Kuyper, and B.A. Thijssen. Competing crossovers in an adaptive GA framework. In *Proceedings of the 1998 IEEE Congress on Evolutionary Computation (CEC 1998)*, pages 787–792. IEEE Press, 1998.
149. A.E. Eiben and J.K. van der Hauw. Solving 3-SAT with adaptive genetic algorithms. In ICEC-97 [231], pages 81–86.

150. A.E. Eiben and J.K. van der Hauw. Graph colouring with adaptive genetic algorithms. *J. Heuristics*, 4:1, 1998.
151. A.E. Eiben and J.I. van Hemert. SAW-ing EAs: Adapting the fitness function for solving constrained problems. In D. Corne, M. Dorigo, and F. Glover, editors, *New Ideas in Optimization*, pages 389–402. McGraw-Hill, 1999.
152. A.E. Eiben, C.H.M. van Kemenade, and J.N. Kok. Orgy in the computer: Multi-parent reproduction in genetic algorithms. In Morán et al. [305], pages 934–945.
153. M.A. El-Beltagy, P.B. Nair, and A.J. Keane. Metamodeling techniques for evolutionary optimization of computationally expensive problems: Promises and limitations. In Banzhaf et al. [36], pages 196–203.
154. N. Eldredge and S.J. Gould. *Models of Paleobiology*, chapter Punctuated Equilibria: an alternative to phyletic gradualism, pages 82–115. Freeman Cooper, San Francisco, 1972.
155. J.M. Epstein and R. Axtell. *Growing Artificial Societies: Social Sciences from Bottom Up*. Brookings Institution Press and The MIT Press, 1996.
156. L.J. Eshelman, editor. *Proceedings of the 6th International Conference on Genetic Algorithms*. Morgan Kaufmann, San Francisco, 1995.
157. L.J. Eshelman, R.A. Caruana, and J.D. Schaffer. Biases in the crossover landscape. In Schaffer [365], pages 10–19.
158. L.J. Eshelman and J.D. Schaffer. Preventing premature convergence in genetic algorithms by preventing incest. In Belew and Booker [46], pages 115–122.
159. L.J. Eshelman and J.D. Schaffer. Crossover's niche. In Forrest [176], pages 9–14.
160. L.J. Eshelman and J.D. Schaffer. Real-coded genetic algorithms and interval schemata. In Whitley [457], pages 187–202.
161. L.J. Eshelman and J.D. Schaffer. Productive recombination and propagating and preserving schemata. In Whitley and Vose [462], pages 299–313.
162. Álvaro Fialho. *Adaptive Operator Selection for Optimization*. PhD thesis, Université Paris-Sud XI, Orsay, France, December 2010.
163. D. Floreano, P. Husbands, and S. Nolfi. Evolutionary robotics. In B. Siciliano and O. Khatib, editors, *Springer Handbook of Robotics*, pages 1423–1451. Springer, 2008.
164. T.C. Fogarty, F. Vavak, and P. Cheng. Use of the genetic algorithm for load balancing of sugar beet presses. In Eshelman [156], pages 617–624.
165. D.B. Fogel. *Evolving Artificial Intelligence*. PhD thesis, University of California, 1992.
166. D.B. Fogel. *Evolutionary Computation*. IEEE Press, 1995.
167. D.B. Fogel, editor. *Evolutionary Computation: the Fossil Record*. IEEE Press, Piscataway, NJ, 1998.
168. D.B. Fogel. *Blondie24: Playing at the Edge of AI*. Morgan Kaufmann, San Francisco, 2002.
169. D.B. Fogel. Better than Samuel: Evolving a nearly expert checkers player. In Ghosh and Tsutsui [184], pages 989–1004.
170. D.B. Fogel and J.W. Atmar. Comparing genetic operators with Gaussian mutations in simulated evolutionary processes using linear systems. *Biological Cybernetics*, 63(2):111–114, 1990.
171. D.B. Fogel and L.C. Stayton. On the effectiveness of crossover in simulated evolutionary optimization. *BioSystems*, 32(3):171–182, 1994.

172. L.J. Fogel, P.J. Angeline, and T. Bäck, editors. *Proceedings of the 5th Annual Conference on Evolutionary Programming*. MIT Press, Cambridge, MA, 1996.
173. L.J. Fogel, A.J. Owens, and M.J. Walsh. Artificial intelligence through a simulation of evolution. In A. Callahan, M. Maxfield, and L.J. Fogel, editors, *Biophysics and Cybernetic Systems*, pages 131–156. Spartan, Washington DC, 1965.
174. L.J. Fogel, A.J. Owens, and M.J. Walsh. *Artificial Intelligence through Simulated Evolution*. Wiley, Chichester, UK, 1966.
175. C.M. Fonseca and P.J. Fleming. Genetic algorithms for multiobjective optimization: formulation, discussion and generalization. In Forrest [176], pages 416–423.
176. S. Forrest, editor. *Proceedings of the 5th International Conference on Genetic Algorithms*. Morgan Kaufmann, San Francisco, 1993.
177. S. Forrest and M. Mitchell. Relative building block fitness and the building block hypothesis. In Whitley [457], pages 109–126.
178. B. Freisleben and P. Merz. A genetic local search algorithm for solving the symmetric and asymetric travelling salesman problem. In ICEC-96 [230], pages 616–621.
179. A.A. Freitas. *Data Mining and Knowledge Discovery with Evolutionary Algorithms*. Springer, 2002.
180. M. Garey and D. Johnson. *Computers and Intractability. A Guide to the Theory of NP-Completeness*. Freeman, San Francisco, 1979.
181. C. Gathercole and P. Ross. Dynamic training subset selection for supervised learning in genetic programming. In Davidor et al. [96], pages 312–321. Lecture Notes in Computer Science 866.
182. D.K. Gehlhaar and D.B. Fogel. Tuning evolutionary programming for conformationally flexible molecular docking. In Fogel et al. [172], pages 419–429.
183. I. Gent and T. Walsh. Phase transitions from real computational problems. In *Proceedings of the 8th International Symposium on Artificial Intelligence*, pages 356–364, 1995.
184. A. Ghosh and S. Tsutsui, editors. *Advances in Evolutionary Computation: Theory and Applications*. Springer, 2003.
185. F. Glover. Tabu search: 1. *ORSA Journal on Computing*, 1(3):190–206, Summer 1989.
186. F. Glover. Tabu search and adaptive memory programming — advances, applications, and challenges. In R.S. Barr, R.V. Helgason, and J.L. Kennington, editors, *Interfaces in Computer Science and Operations Research*, pages 1–75. Kluwer Academic Publishers, Norwell, MA, 1996.
187. D.E. Goldberg. Genetic algorithms and Walsh functions: I. A gentle introduction. *Complex Systems*, 3(2):129–152, April 1989.
188. D.E. Goldberg. Genetic algorithms and Walsh functions: II. Deception and its analysis. *Complex Systems*, 3(2):153–171, April 1989.
189. D.E. Goldberg. *Genetic Algorithms in Search, Optimization and Machine Learning*. Addison-Wesley, 1989.
190. D.E. Goldberg and K. Deb. A comparative analysis of selection schemes used in genetic algorithms. In Rawlins [351], pages 69–93.
191. D.E. Goldberg, B. Korb, and K. Deb. Messy genetic algorithms: Motivation, analysis, and first results. *Complex Systems*, 3(5):493–530, October 1989.
192. D.E. Goldberg and R. Lingle. Alleles, loci, and the traveling salesman problem. In Grefenstette [197], pages 154–159.

193. D.E. Goldberg and J. Richardson. Genetic algorithms with sharing for multimodal function optimization. In Grefenstette [198], pages 41–49.
194. D.E. Goldberg and R.E. Smith. Nonstationary function optimization using genetic algorithms with dominance and diploidy. In Grefenstette [198], pages 59–68.
195. M. Gorges-Schleuter. ASPARAGOS: An asynchronous parallel genetic optimization strategy. In Schaffer [365], pages 422–427.
196. J. Gottlieb and G.R. Raidl. The effects of locality on the dynamics of decoder-based evolutionary search. In Whitley et al. [453], pages 283–290.
197. J.J. Grefenstette, editor. *Proceedings of the 1st International Conference on Genetic Algorithms and Their Applications*. Lawrence Erlbaum, Hillsdale, New Jersey, 1985.
198. J.J. Grefenstette, editor. *Proceedings of the 2nd International Conference on Genetic Algorithms and Their Applications*. Lawrence Erlbaum, Hillsdale, New Jersey, 1987.
199. J.J. Grefenstette. Genetic algorithms for changing environments. In Männer and Manderick [282], pages 137–144.
200. J.J. Grefenstette. Deception considered harmful. In Whitley [457], pages 75–91.
201. J.J. Grefenstette, R. Gopal, B. Rosmaita, and D. van Guch. Genetic algorithm for the TSP. In Grefenstette [197], pages 160–168.
202. J.J Greffenstette. Optimisation of Control Parameters for Genetic Algorithms. *IEEE Transactions on Systems, Man and Cybernetics*, 16(1):122–128, 1986.
203. J.J. Merelo Guervos, P. Adamidis, H.-G. Beyer, J.-L. Fernandez-Villacanas, and H.-P. Schwefel, editors. *Proceedings of the 7th Conference on Parallel Problem Solving from Nature*, number 2439 in Lecture Notes in Computer Science. Springer, 2002.
204. E. Haasdijk, N. Bredeche, and A. E. Eiben. Combining environment-driven adaptation and task-driven optimisation in evolutionary robotics. *PLoS ONE*, 9(6):e98466, 2014.
205. A.B. Hadj-Alouane and J.C. Bean. A genetic algorithm for the multiple-choice integer program. Technical Report 92-50, University of Michigan, 1992.
206. N. Hansen. The CMA evolution strategy: A comparing review. In J.A. Lozano and P. Larranaga, editors, *Towards a New Evolutionary Computation : Advances in Estimation of Distribution Algorithms*, pages 75–102. Springer, 2006.
207. N. Hansen and A. Ostermeier. Completely derandomized self-adaptation in evolution strategies. *Evolutionary Computation*, 9(2):159–195, 2001.
208. P. Hansen and N. Mladenoviè. An introduction to variable neighborhood search. In S. Voß, S. Martello, I.H. Osman, and C. Roucairol, editors, *Meta-heuristics: Advances and Trends in Local Search Paradigms for Optimization. Proceedings of MIC 97 Conference*. Kluwer Academic Publishers, Dordrecht, The Netherlands, 1998.
209. G. Harik and D.E. Goldberg. Learning linkage. In R.K. Belew and M.D. Vose, editors, *Foundations of Genetic Algorithms 4*, pages 247–262. Morgan Kaufmann, San Francisco, 1996.
210. E. Hart, P. Ross, and J. Nelson. Solving a real-world problem using an evolving heuristically driven schedule builder. *Evolutionary Computation*, 6(1):61–81, 1998.
211. W.E. Hart. *Adaptive Global Optimization with Local Search*. PhD thesis, University of California, San Diego, 1994.

212. M.S. Hillier and F.S. Hillier. Conventional optimization techniques. In Sarker et al. [363], chapter 1, pages 3–25.
213. W.D. Hillis. Co-evolving parasites improve simulated evolution as an optimization procedure. In C.G. Langton, C. Taylor, J.D.. Farmer, and S. Rasmussen, editors, *Proceedings of the Workshop on Artificial Life (ALIFE '90)*, pages 313–324, Redwood City, CA, USA, 1992. Addison-Wesley.
214. R. Hinterding, Z. Michalewicz, and A.E. Eiben. Adaptation in evolutionary computation: A survey. In ICEC-97 [231].
215. G.E. Hinton and S.J. Nowlan. How learning can guide evolution. *Complex Systems*, 1:495–502, 1987.
216. P.G. Hoel, S.C. Port, and C.J. Stone. *Introduction to Stochastic Processes.* Houghton Mifflin, 1972.
217. F. Hoffmeister and T. Bäck. Genetic self-learning. In Varela and Bourgine [440], pages 227–235.
218. J.H. Holland. Genetic algorithms and the optimal allocation of trials. *SIAM J. of Computing*, 2:88–105, 1973.
219. J.H. Holland. Adaptation. In Rosen and Snell, editors, *Progress in Theoretical Biology: 4.* Plenum, 1976.
220. J.H. Holland. *Adaption in Natural and Artificial Systems.* MIT Press, Cambridge, MA, 1992. 1st edition: 1975, The University of Michigan Press, Ann Arbor.
221. J.N. Hooker. Testing heuristics: We have it all wrong. *Journal of Heuristics*, 1:33–42, 1995.
222. W. Hordijk and B. Manderick. The usefulness of recombination. In Morán et al. [305], pages 908–919.
223. J. Horn, N. Nafpliotis, and D.E. Goldberg. A niched Pareto genetic algorithm for multiobjective optimization. In ICEC-94 [229], pages 82–87.
224. C.R. Houck, J.A. Joines, M.G. Kay, and J.R. Wilson. Empirical investigation of the benefits of partial Lamarckianism. *Evolutionary Computation*, 5(1):31–60, 1997.
225. P. Husbands. Distributed co-evolutionary genetic algorithms for multi-criteria and multi-constraint optimisiation. In T.C. Fogarty, editor, *Evolutionary Computing: Proceedings of the AISB workshop*, LNCS 865, pages 150–165. Springer, 1994.
226. F. Hutter, T. Bartz-Beielstein, H.H. Hoos, K. Leyton-Brown, and K.P. Murphy. Sequential model-based parameter optimisation: an experimental investigation of automated and interactive approaches. In T. Bartz-Beielstein, M. Chiarandini, L. Paquete, and M. Preuss, editors, *Empirical Methods for the Analysis of Optimization Algorithms*, chapter 15, pages 361–411. Springer, 2010.
227. F. Hutter, H.H. Hoos, K. Leyton-Brown, and T. Stützle. ParamILS: an automatic algorithm configuration framework. *Journal of Artificial Intelligence Research*, 36:267–306, October 2009.
228. H. Iba, H. de Garis, and T. Sato. Genetic programming using a minimum description length principle. In Kinnear [249], pages 265–284.
229. *Proceedings of the First IEEE Conference on Evolutionary Computation.* IEEE Press, Piscataway, NJ, 1994.
230. *Proceedings of the 1996 IEEE Conference on Evolutionary Computation.* IEEE Press, Piscataway, NJ, 1996.
231. *Proceedings of the 1997 IEEE Conference on Evolutionary Computation.* IEEE Press, Piscataway, NJ, 1997.

232. A. Jain and D.B. Fogel. Case studies in applying fitness distributions in evolutionary algorithms. II. Comparing the improvements from crossover and gaussian mutation on simple neural networks. In X. Yao and D.B. Fogel, editors, *Proc. of the 2000 IEEE Symposium on Combinations of Evolutionary Computation and Neural Networks*, pages 91–97. IEEE Press, Piscataway, NJ, 2000.
233. N. Jakobi, P. Husbands, and I. Harvey. Noise and the reality gap: The use of noise and the reality gap: The use of simulation in evolutionary robotics. In *Proc. of the Third European Conference on Artificial Life*, number 929 in LNCS, pages 704–720. Springer, 1995.
234. T. Jansen. *Analyzing Evolutionary Algorithms. The Computer Science Perspective.* Natural Computing Series. Springer, 2013.
235. Y. Jin. A comprehensive survey of fitness approximation in evolutionary computation. *Soft Computing*, 9(1):3–12, 2005.
236. Y. Jin. Surrogate-assisted evolutionary computation: Recent advances and future challenges. *Swarm and Evolutionary Computation*, 1(2):61–70, 2011.
237. J.A. Joines and C.R. Houck. On the use of non-stationary penalty functions to solve nonlinear constrained optimisation problems with GA's. In ICEC-94 [229], pages 579–584.
238. T. Jones. *Evolutionary Algorithms, Fitness Landscapes and Search.* PhD thesis, University of New Mexico, Albuquerque, NM, 1995.
239. L. Kallel, B. Naudts, and C. Reeves. Properties of fitness functions and search landscapes. In Kallel et al. [240], pages 175–206.
240. L. Kallel, B. Naudts, and A. Rogers, editors. *Theoretical Aspects of Evolutionary Computing.* Springer, 2001.
241. G. Karafotias, M. Hoogendoorn, and A.E. Eiben. Trends and challenges in evolutionary algorithms parameter control. *IEEE Transactions on Evolutionary Computation*, 19(2):167–187, 2015.
242. G. Karafotias, S.K. Smit, and A.E. Eiben. A generic approach to parameter control. In C. Di Chio et al., editor, *Applications of Evolutionary Computing, EvoStar 2012*, volume 7248 of *LNCS*, pages 361–370. Springer, 2012.
243. H. Kargupta. The gene expression messy genetic algorithm. In ICEC-96 [230], pages 814–819.
244. S.A. Kauffman. *Origins of Order: Self-Organization and Selection in Evolution.* Oxford University Press, New York, NY, 1993.
245. A.J. Keane and S.M. Brown. The design of a satellite boom with enhanced vibration performance using genetic algorithm techniques. In I.C. Parmee, editor, *Proceedings of the Conference on Adaptive Computing in Engineering Design and Control 96*, pages 107–113. P.E.D.C., Plymouth, 1996.
246. G. Kendall, P. Cowling, and E. Soubeiga. Choice function and random hyperheuristics. In *Proceedings of Fourth Asia-Pacific Conference on Simulated Evolution and Learning (SEAL)*, pages 667–671, 2002.
247. J. Kennedy and R. Eberhart. Particle swarm optimization. In *Proceedings of the 1995 IEEE International Conference on Neural Networks*, volume 4, pages 1942–1948, November 1995.
248. J. Kennedy and R.C. Eberhart. *Swarm Intelligence.* Morgan Kaufmann, 2001.
249. K.E. Kinnear, editor. *Advances in Genetic Programming.* MIT Press, Cambridge, MA, 1994.
250. S. Kirkpatrick, C. Gelatt, and M. Vecchi. Optimization by simulated anealing. *Science*, 220:671–680, 1983.

251. J.D. Knowles and D.W. Corne. Approximating the nondominated front using the Pareto Archived Evolution Strategy. *Evolutionary Computation*, 8(2):149–172, 2000.
252. J.R. Koza. *Genetic Programming.* MIT Press, Cambridge, MA, 1992.
253. J.R. Koza. *Genetic Programming II.* MIT Press, Cambridge, MA, 1994.
254. J.R. Koza. Scalable learning in genetic programming using automatic function definition. In Kinnear [249], pages 99–117.
255. J.R. Koza and F.H. Bennett. Automatic synthesis, placement, and routing of electrical circuits by means of genetic programming. In Spector et al. [416], pages 105–134.
256. J.R. Koza, K. Deb, M. Dorigo, D.B. Fogel, M. Garzon, H. Iba, and R.L. Riolo, editors. *Proceedings of the 2nd Annual Conference on Genetic Programming.* MIT Press, Cambridge, MA, 1997.
257. Oliver Kramer. Evolutionary self-adaptation: a survey of operators and strategy parameters. *Evolutionary Intelligence*, 3(2):51–65, 2010.
258. N. Krasnogor. Coevolution of genes and memes in memetic algorithms. In A.S. Wu, editor, *Proceedings of the 1999 Genetic and Evolutionary Computation Conference Workshop Program*, 1999.
259. N. Krasnogor. *Studies in the Theory and Design Space of Memetic Algorithms.* PhD thesis, University of the West of England, 2002.
260. N. Krasnogor. Self-generating metaheuristics in bioinformatics: The protein structure comparison case. *Genetic Programming and Evolvable Machines. Kluwer academic Publishers*, 5(2):181–201, 2004.
261. N. Krasnogor, B.P. Blackburne, E.K. Burke, and J.D. Hirst. Multimeme algorithms for protein structure prediction. In Guervos et al. [203], pages 769–778.
262. N. Krasnogor and S.M. Gustafson. A study on the use of "self-generation" in memetic algorithms. *Natural Computing*, 3(1):53–76, 2004.
263. N. Krasnogor and J.E. Smith. A memetic algorithm with self-adaptive local search: TSP as a case study. In Whitley et al. [453], pages 987–994.
264. N. Krasnogor and J.E. Smith. Emergence of profitable search strategies based on a simple inheritance mechanism. In Spector et al. [415], pages 432–439.
265. N. Krasnogor and J.E. Smith. A tutorial for competent memetic algorithms: Model, taxonomy and design issues. *IEEE Transactions on Evolutionary Computation*, 9(5):474–488, 2005.
266. T. Krink, P. Rickers, and R. Thomsen. Applying self-organised criticality to evolutionary algorithms. In Schoenauer et al. [368], pages 375–384.
267. M.W.S. Land. *Evolutionary Algorithms with Local Search for Combinatorial Optimization.* PhD thesis, University of California, San Diego, 1998.
268. W.B. Langdon, T. Soule, R. Poli, and J.A. Foster. The evolution of size and shape. In Spector et al. [416], pages 163–190.
269. P.L. Lanzi. Learning classifier systems: then and now. *Evolutionary Intelligence*, 1:63–82, 2008.
270. P.L. Lanzi, W. Stolzmann, and S.W. Wilson, editors. *Learning Classifier Systems: From Foundations to Applications*, volume 1813 of *LNAI*. Springer, 2000.
271. S. Lin and B. Kernighan. An effective heuristic algorithm for the Traveling Salesman Problem. *Operations Research*, 21:498–516, 1973.
272. X. Llora, R. Reddy, B. Matesic, and R. Bhargava. Towards better than human capability in diagnosing prostate cancer using infrared spectroscopic imaging. In Bosman et al. [65], pages 2098–2105.

273. F.G. Lobo, C.F. Lima, and Z. Michalewicz, editors. *Parameter Setting in Evolutionary Algorithms*. Springer, 2007.
274. R. Lohmann. Application of evolution strategy in parallel populations. In Schwefel and Männer [374], pages 198–208.
275. S. Luke and L. Spector. A comparison of crossover and mutation in genetic programming. In Koza et al. [256], pages 240–248.
276. G. Luque and E. Alba. *Parallel Genetic Algorithms*, volume 367 of *Studies in Computational Intelligence*. Springer, 2011.
277. W.G. Macready and D.H. Wolpert. Bandit problems and the exploration/exploitation tradeoff. *IEEE Transactions on Evolutionary Computation*, 2(1):2–22, April 1998.
278. S.W. Mahfoud. Crowding and preselection revisited. In Männer and Manderick [282], pages 27–36.
279. S.W. Mahfoud. Boltzmann selection. In Bäck et al. [26], pages C2.5:1–4.
280. R. Mallipeddi and P. Suganthan. Differential evolution algorithm with ensemble of parameters and mutation and crossover strategies. In B. Panigrahi, S. Das, P. Suganthan, and S. Dash, editors, *Swarm, Evolutionary, and Memetic Computing*, volume 6466 of *Lecture Notes in Computer Science*, pages 71–78. Springer, 2010.
281. B. Manderick and P. Spiessens. Fine-grained parallel genetic algorithms. In Schaffer [365], pages 428–433.
282. R. Männer and B. Manderick, editors. *Proceedings of the 2nd Conference on Parallel Problem Solving from Nature*. North-Holland, Amsterdam, 1992.
283. O. Maron and A. Moore. The racing algorithm: Model selection for lazy learners. In *Artificial Intelligence Review*, volume 11, pages 193–225. Kluwer Academic Publishers, USA, April 1997.
284. W.N. Martin, J. Lienig, and J.P. Cohoon. Island (migration) models: evolutionary algorithms based on punctuated equilibria. In Bäck et al. [28], chapter 15, pages 101–124.
285. W.N. Martin and W.M. Spears, editors. *Foundations of Genetic Algorithms 6*. Morgan Kaufmann, San Francisco, 2001.
286. J. Maturana, F. Lardeux, and F. Saubion. Autonomous operator management for evolutionary algorithms. *Journal of Heuristics*, 16:881–909, 2010.
287. G. Mayley. Landscapes, learning costs and genetic assimilation. *Evolutionary Computation*, 4(3):213–234, 1996.
288. J. Maynard-Smith. *The Evolution of Sex*. Cambridge University Press, Cambridge, UK, 1978.
289. J. Maynard-Smith and E. Száthmary. *The Major Transitions in Evolution*. W.H. Freeman, 1995.
290. J.T. McClave and T. Sincich. *Statistics*. Prentice Hall, 9th edition, 2003.
291. B. McGinley, J. Maher, C. O'Riordan, and F. Morgan. Maintaining healthy population diversity using adaptive crossover, mutation, and selection. *IEEE Transactions on Evolutionary Computation*, 15(5):692 –714, 2011.
292. P. Merz. *Memetic Algorithms for Combinatorial Optimization Problems: Fitness Landscapes and Effective Search Strategies*. PhD thesis, University of Siegen, Germany, 2000.
293. P. Merz and B. Freisleben. Fitness landscapes and memetic algorithm design. In D. Corne, M. Dorigo, and F. Glover, editors, *New Ideas in Optimization*, pages 245–260. McGraw Hill, London, 1999.

294. R. Meuth, M.H. Lim, Y.S. Ong, and D.C. Wunsch. A proposition on memes and meta-memes in computing for higher-order learning. *Memetic Computing*, 1(2):85–100, 2009.
295. Z. Michalewicz. *Genetic Algorithms + Data Structures = Evolution Programs*. Springer, 3rd edition, 1996.
296. Z. Michalewicz. *Genetic algorithms + data structures = evolution programs (3nd, extended ed.)*. Springer, New York, NY, USA, 1996.
297. Z. Michalewicz. Decoders. In Bäck et al. [28], chapter 8, pages 49–55.
298. Z. Michalewicz, K. Deb, M. Schmidt, and T. Stidsen. Test-case generator for nonlinear continuous parameter optimization techniques. *IEEE Transactions on Evolutionary Computation*, 4(3):197–215, 2000.
299. Z. Michalewicz and G. Nazhiyath. Genocop III: A coevolutionary algorithm for numerical optimisation problems with nonlinear constraintrs. In *Proceedings of the 1995 IEEE Conference on Evolutionary Computation*, pages 647–651. IEEE Press, Piscataway, NJ, 1995.
300. Z. Michalewicz and M. Schmidt. Evolutionary algorithms and constrained optimization. In Sarker et al. [363], chapter 3, pages 57–86.
301. Z. Michalewicz and M. Schmidt. TCG-2: A test-case generator for nonlinear parameter optimisation techniques. In Ghosh and Tsutsui [184], pages 193–212.
302. Z. Michalewicz and M. Schoenauer. Evolutionary algorithms for constrained parameter optimisation problems. *Evolutionary Computation*, 4(1):1–32, 1996.
303. J.F. Miller, T. Kalganova, D. Job, and N. Lipnitskaya. The genetic algorithm as a discovery engine: Strange circuits and new principles. In Bentley and Corne [49], pages 443–466.
304. D.J. Montana. Strongly typed genetic programming. *Evolutionary Computation*, 3(2):199–230, 1995.
305. F. Morán, A. Moreno, J. J. Merelo, and P. Chacón, editors. *Advances in Artificial Life. Third International Conference on Artificial Life*, volume 929 of *Lecture Notes in Artificial Intelligence*. Springer, 1995.
306. N. Mori, H. Kita, and Y. Nishikawa. Adaptation to a changing environment by means of the thermodynamical genetic algorithm. In Voigt et al. [445], pages 513–522.
307. A. Moroni, J. Manzolli, F. Von Zuben, and R. Gudwin. Vox populi: Evolutionary computation for music evolution. In Bentley and Corne [49], pages 206–221.
308. P.A. Moscato. On evolution, search, optimization, genetic algorithms and martial arts: Towards memetic algorithms. Technical Report Caltech Concurrent Computation Program Report 826, Caltech, 1989.
309. P.A. Moscato. *Problemas de Otimizacão NP, Aproximabilidade e Computacão Evolutiva: Da Prática à Teoria*. PhD thesis, Universidade Estadual de Campinas, Brazil, 2001.
310. T. Motoki. Calculating the expected loss of diversity of selection schemes. *Evolutionary Computation*, 10(4):397–422, 2002.
311. H. Mühlenbein. Parallel genetic algorithms, population genetics and combinatorial optimization. In Schaffer [365], pages 416–421.
312. M. Munetomo and D.E. Goldberg. Linkage identification by non-monotonicity detection for overlapping functions. *Evolutionary Computation*, 7(4):377–398, 1999.

313. V. Nannen and A. E. Eiben. A method for parameter calibration and relevance estimation in evolutionary algorithms. In M. Keijzer, editor, *Proceedings of the Genetic and Evolutionary Computation Conference (GECCO-2006)*, pages 183–190. Morgan Kaufmann, San Francisco, 2006.
314. V. Nannen and A. E. Eiben. Relevance Estimation and Value Calibration of Evolutionary Algorithm Parameters. In Manuela M. Veloso, editor, *Proceedings of the 20th International Joint Conference on Artificial Intelligence (IJCAI)*, pages 1034–1039. Hyderabad, India, 2007.
315. A.L. Nelson, G.J. Barlow, and L. Doitsidis. Fitness functions in evolutionary robotics: A survey and analysis. *Robotics and Autonomous Systems*, 57(4):345–370, 2009.
316. F. Neri. An Adaptive Multimeme Algorithm for Designing HIV Multidrug Therapies. *IEEE/ACM Transactions on Computational Biology and Bioinformatics*, 4(2):264–278, 2007.
317. F. Neri. Fitness diversity based adaptation in Multimeme Algorithms: A comparative study. In *IEEE Congress on Evolutionary Computation, CEC 2007*, pages 2374–2381, 2007.
318. F. Neumann and C. Witt. *Bioinspired Computation in Combinatorial Optimization: Algorithms and Their Computational Complexity.* Natural Computing Series. Springer, 2010.
319. Q.H. Nguyen, Y.-S. Ong, M.H. Lim, and N. Krasnogor. Adaptive cellular memetic algorithms. *Evolutionary Computation*, 17(2), June 2009.
320. Q.H. Nguyen, Y.S. Ong, and M.H. Lim. A Probabilistic Memetic Framework. *IEEE Transactions on Evolutionary Computation*, 13(3):604–623, 2009.
321. A. Nix and M. Vose. Modelling genetic algorithms with Markov chains. *Annals of Mathematics and Artifical Intelligence*, pages 79–88, 1992.
322. S. Nolfi and D. Floreano. *Evolutionary Robotics: The Biology, Intelligence, and Technology of Self-Organizing Machines.* MIT Press, Cambridge, MA, 2000.
323. P. O'Dowd, A.F. Winfield, and M. Studley. The distributed co-evolution of an embodied simulator and controller for swarm robot behaviours. In *IEEE/RSJ International Conference on Intelligent Robots and Systems (IROS 2011)*, pages 4995–5000. IEEE Press, 2011.
324. C.K. Oei, D.E. Goldberg, and S.J. Chang. Tournament selection, niching, and the preservation of diversity. Technical Report 91011, University of Illinois Genetic Algorithms Laboratory, 1991.
325. I.M. Oliver, D.J. Smith, and J. Holland. A study of permutation crossover operators on the travelling salesman problem. In Grefenstette [198], pages 224–230.
326. Y.S. Ong and A.J. Keane. Meta-lamarckian learning in memetic algorithms. *IEEE Transactions on Evolutionary Computation*, 8(2):99–110, 2004.
327. Y.S. Ong, M.H. Lim, and X. Chen. Memetic Computation—Past, Present & Future [Research Frontier]. *Computational Intelligence Magazine, IEEE*, 5(2):24–31, 2010.
328. Y.S. Ong, M.H. Lim, N. Zhu, and K.W. Wong. Classification of adaptive memetic algorithms: A comparative study. *IEEE Transactions on Systems Man and Cybernetics Part B*, 36(1), 2006.
329. B. Paechter, R.C. Rankin, A. Cumming, and T.C. Fogarty. Timetabling the classes of an entire university with an evolutionary algorithm. In Eiben et al. [131], pages 865–874.

330. C.M. Papadimitriou. *Computational complexity.* Addison-Wesley, Reading, Massachusetts, 1994.
331. C.M. Papadimitriou and K. Steiglitz. *Combinatorial optimization: algorithms and complexity.* Prentice Hall, Englewood Cliffs, NJ, 1982.
332. J. Paredis. The symbiotic evolution of solutions and their representations. In Eshelman [156], pages 359–365.
333. J. Paredis. Coevolutionary algorithms. In Bäck et al. [26].
334. I.C. Parmee. Improving problem definition through interactive evolutionary computing. *Artificial Intelligence for Engineering Design, Analysis and Manufacturing*, 16(3):185–202, 2002.
335. O. Pauplin, P. Caleb-Solly, and J.E. Smith. User-centric image segmentation using an interactive parameter adaptation tool. *Pattern Recognition*, 43(2):519–529, February 2010.
336. M. Pelikan, D.E. Goldberg, and E. Cantù-Paz. BOA: The Bayesian optimization algorithm. In Banzhaf et al. [36], pages 525–532.
337. M. Pelikan and H. Mühlenbein. The bivariate marginal distribution algorithm. In R. Roy, T. Furuhashi, and P.K. Chawdhry, editors, *Advances in Soft Computing–Engineering Design and Manufacturing*, pages 521–535. Springer, 1999.
338. C. Pettey. Diffusion (cellular) models. In Bäck et al. [28], chapter 16, pages 125–133.
339. C.B. Pettey, M.R. Leuze, and J.J. Grefenstette. A parallel genetic algorithm. In Grefenstette [198], pages 155–161.
340. R. Poli, J. Kennedy, and T. Blackwell. Particle swarm optimization – an overview. *Swarm Intelligence*, 1(1):33–57, 2007.
341. C. Potta, R. Poli, J. Rowe, and K. De Jong, editors. *Foundations of Genetic Algorithms 7.* Morgan Kaufmann, San Francisco, 2003.
342. M.A. Potter and K.A. De Jong. A cooperative coevolutionary approach to function optimisation. In Davidor et al. [96], pages 248–257.
343. K.V. Price, R.N. Storn, and J.A. Lampinen. *Differential Evolution: A Practical Approach to Global Optimization.* Natural Computing Series. Springer, 2005.
344. P. Prosser. An empirical study of phase transitions in binary constraint satisfaction problems. *Artificial Intelligence*, 81:81–109, 1996.
345. A. Prügel-Bennet and J. Shapiro. An analysis of genetic algorithms using statistical mechanics. *Phys. Review Letters*, 72(9):1305–1309, 1994.
346. A. Prügel-Bennett. Modelling evolving populations. *J. Theoretical Biology*, 185(1):81–95, March 1997.
347. A. Prügel-Bennett. Modelling finite populations. In Potta et al. [341].
348. A. Prügel-Bennett and A. Rogers. Modelling genetic algorithm dynamics. In L. Kallel, B. Naudts, and A. Rogers, editors, *Theoretical Aspects of Evolutionary Computing*, pages 59–85. Springer, 2001.
349. A. K. Qin, V. L. Huang, and P. N. Suganthan. Differential evolution algorithm with strategy adaptation for global numerical optimization. *Trans. Evol. Comp*, 13:398–417, April 2009.
350. N. Radcliffe. Forma analysis and random respectful recombination. In Belew and Booker [46], pages 222–229.
351. G. Rawlins, editor. *Foundations of Genetic Algorithms.* Morgan Kaufmann, San Francisco, 1991.

352. I. Rechenberg. *Evolutionstrategie: Optimierung technisher Systeme nach Prinzipien des biologischen Evolution.* Frommann-Hollboog Verlag, Stuttgart, 1973.
353. C. Reeves and J. Rowe. *Genetic Algorithms: Principles and Perspectives.* Kluwer, Norwell MA, 2002.
354. A. Rogers and A. Prügel-Bennett. Modelling the dynamics of a steady-state genetic algorithm. In Banzhaf and Reeves [38], pages 57–68.
355. J. Romero and P. Machado. *The Art of Artificial Evolution.* Natural Computing Series. Springer, 2008.
356. G. Rudolph. Global optimization by means of distributed evolution strategies. In Schwefel and Männer [374], pages 209–213.
357. G. Rudolph. Convergence properties of canonical genetic algorithms. *IEEE Transactions on Neural Networks*, 5(1):96–101, 1994.
358. G. Rudolph. Convergence of evolutionary algorithms in general search spaces. In ICEC-96 [230], pages 50–54.
359. G. Rudolph. Reflections on bandit problems and selection methods in uncertain environments. In Bäck [23], pages 166–173.
360. G. Rudolph. Takeover times and probabilities of non-generational selection rules. In Whitley et al. [453], pages 903–910.
361. T. Runarson and X. Yao. Constrained evolutionary optimization – the penalty function approach. In Sarker et al. [363], chapter 4, pages 87–113.
362. A. Salman, K. Mehrota, and C. Mohan. Linkage crossover for genetic algorithms. In Banzhaf et al. [36], pages 564–571.
363. R. Sarker, M. Mohammadian, and X. Yao, editors. *Evolutionary Optimization.* Kluwer Academic Publishers, Boston, 2002.
364. J.D. Schaffer. *Multiple Objective Optimization with Vector Evaluated Genetic Algorithms.* PhD thesis, Vanderbilt University, Tennessee, 1984.
365. J.D. Schaffer, editor. *Proceedings of the 3rd International Conference on Genetic Algorithms.* Morgan Kaufmann, San Francisco, 1989.
366. J.D. Schaffer and L.J. Eshelman. On crossover as an evolutionarily viable strategy. In Belew and Booker [46], pages 61–68.
367. D. Schlierkamp-Voosen and H. Mühlenbein. Strategy adaptation by competing subpopulations. In Davidor et al. [96], pages 199–209.
368. M. Schoenauer, K. Deb, G. Rudolph, X. Yao, E. Lutton, J.J. Merelo, and H.-P. Schwefel, editors. *Proceedings of the 6th Conference on Parallel Problem Solving from Nature*, number 1917 in Lecture Notes in Computer Science. Springer, 2000.
369. M. Schoenauer and S. Xanthakis. Constrained GA optimisation. In Forrest [176], pages 573–580.
370. S. Schulenburg and P. Ross. Strength and money: An LCS approach to increasing returns. In P.L. Lanzi, W. Stolzmann, and S.W. Wilson, editors, *Advances in Learning Classifier Systems*, volume 1996 of *LNAI*, pages 114–137. Springer, 2001.
371. M.C. Schut, E. Haasdijk, and A.E. Eiben. What is situated evolution? In *Proceedings of the 2009 IEEE Congress on Evolutionary Computation (CEC 2009)*, pages 3277–3284. IEEE Press, 2009.
372. H.-P. Schwefel. *Numerische Optimierung von Computer-Modellen mittels der Evolutionsstrategie*, volume 26 of *ISR*. Birkhaeuser, Basel/Stuttgart, 1977.
373. H.-P. Schwefel. *Evolution and Optimum Seeking.* Wiley, New York, 1995.

374. H.-P. Schwefel and R. Männer, editors. *Proceedings of the 1st Conference on Parallel Problem Solving from Nature*, number 496 in Lecture Notes in Computer Science. Springer, 1991.
375. M Serpell and JE Smith. Self-Adaption of Mutation Operator and Probability for Permutation Representations in Genetic Algorithms. *Evolutionary Computation*, 18(3):1–24, February 2010.
376. S. K. Smit and A. E. Eiben. Beating the 'world champion' evolutionary algorithm via REVAC tuning. In *IEEE Congress on Evolutionary Computation*, pages 1–8, Barcelona, Spain, 2010. IEEE Computational Intelligence Society, IEEE Press.
377. S.K. Smit. *Algorithm Design in Experimental Research.* Ph.D Thesis, Vrije Universiteit Amsterdam, 2012.
378. S.K. Smit and A.E. Eiben. Comparing parameter tuning methods for evolutionary algorithms. In *Proceedings of the 2009 IEEE Congress on Evolutionary Computation (CEC 2009)*, pages 399–406. IEEE Press, 2009.
379. A.E. Smith and D.W. Coit. Penalty functions. In Bäck et al. [28], chapter 7, pages 41–48.
380. A.E. Smith and D.M. Tate. Genetic optimisation using a penalty function. In Forrest [176], pages 499–505.
381. J. Smith. Credit assignment in adaptive memetic algorithms. In *Proceedings of GECCO 2007, the ACM-SIGEVO conference on Evolutionary Computation*, pages 1412–1419, 2007.
382. J.E. Smith. *Self-Adaptation in Evolutionary Algorithms*. PhD thesis, University of the West of England, Bristol, UK, 1998.
383. J.E. Smith. Modelling GAs with self-adaptive mutation rates. In Spector et al. [415], pages 599–606.
384. J.E. Smith. Co-evolution of memetic algorithms: Initial investigations. In Guervos et al. [203], pages 537–548.
385. J.E. Smith. On appropriate adaptation levels for the learning of gene linkage. *J. Genetic Programming and Evolvable Machines*, 3(2):129–155, 2002.
386. J.E. Smith. Parameter perturbation mechanisms in binary coded GAs with self-adaptive mutation. In Rowe, Poli, DeJong, and Cotta, editors, *Foundations of Genetic Algorithms 7*, pages 329–346. Morgan Kaufmann, San Francisco, 2003.
387. J.E. Smith. Protein structure prediction with co-evolving memetic algorithms. In *Proceedings of the 2003 Congress on Evolutionary Computation (CEC 2003)*, pages 2346–2353. IEEE Press, Piscataway, NJ, 2003.
388. J.E. Smith. The co-evolution of memetic algorithms for protein structure prediction. In W.E. Hart, N. Krasnogor, and J.E. Smith, editors, *Recent Advances in Memetic Algorithms*, pages 105–128. Springer, 2004.
389. J.E. Smith. Co-evolving memetic algorithms: A review and progress report. *IEEE Transactions in Systems, Man and Cybernetics, part B*, 37(1):6–17, 2007.
390. J.E. Smith. On replacement strategies in steady state evolutionary algorithms. *Evolutionary Computation*, 15(1):29–59, 2007.
391. J.E. Smith. Estimating meme fitness in adaptive memetic algorithms for combinatorial problems. *Evolutionary Computation*, 20(2):165188, 2012.
392. J.E. Smith, M. Bartley, and T.C. Fogarty. Microprocessor design verification by two-phase evolution of variable length tests. In ICEC-97 [231], pages 453–458.
393. J.E. Smith and T.C. Fogarty. An adaptive poly-parental recombination strategy. In T.C. Fogarty, editor, *Evolutionary Computing 2*, pages 48–61. Springer, 1995.

394. J.E. Smith and T.C. Fogarty. Evolving software test data - GAs learn self expression. In T.C. Fogarty, editor, *Evolutionary Computing*, number 1143 in Lecture Notes in Computer Science, pages 137–146, 1996.
395. J.E. Smith and T.C. Fogarty. Recombination strategy adaptation via evolution of gene linkage. In ICEC-96 [230], pages 826–831.
396. J.E. Smith and T.C. Fogarty. Self adaptation of mutation rates in a steady state genetic algorithm. In ICEC-96 [230], pages 318–323.
397. J.E. Smith and T.C. Fogarty. Operator and parameter adaptation in genetic algorithms. *Soft Computing*, 1(2):81–87, 1997.
398. J.E. Smith, T.C. Fogarty, and I.R. Johnson. Genetic feature selection for clustering and classification. In *Proceedings of the IEE Colloquium on Genetic Algorithms in Image Processing and Vision*, volume IEE Digest 1994/193, 1994.
399. J.E. Smith and F. Vavak. Replacement strategies in steady state genetic algorithms: dynamic environments. *J. Computing and Information Technology*, 7(1):49–60, 1999.
400. J.E. Smith and F. Vavak. Replacement strategies in steady state genetic algorithms: static environments. In Banzhaf and Reeves [38], pages 219–234.
401. R.E. Smith, C. Bonacina, P. Kearney, and W. Merlat. Embodiment of evolutionary computation in general agents. *Evolutionary Computation*, 8(4):475–493, 2001.
402. R.E. Smith and D.E. Goldberg. Diploidy and dominance in artificial genetic search. *Complex Systems*, 6:251–285, 1992.
403. R.E. Smith and J.E. Smith. An examination of tuneable, random search landscapes. In Banzhaf and Reeves [38], pages 165–181.
404. R.E. Smith and J.E. Smith. New methods for tuneable, random landscapes. In Martin and Spears [285], pages 47–67.
405. C. Soddu. Recognizability of the idea: the evolutionary process of Argenia. In Bentley and Corne [49], pages 109–127.
406. T. Soule and J.A. Foster. Effects of code growth and parsimony pressure on populations in genetic programming. *Evolutionary Computation*, 6(4):293–309, Winter 1998.
407. T. Soule, J.A. Foster, and J. Dickinson. Code growth in genetic programming. In J.R. Koza, D.E. Goldberg, D.B. Fogel, and R.L. Riolo, editors, *Proceedings of the 1st Annual Conference on Genetic Programming*, pages 215–223. MIT Press, Cambridge, MA, 1996.
408. W.M. Spears. Crossover or mutation. In Whitley [457], pages 220–237.
409. W.M. Spears. Simple subpopulation schemes. In A.V. Sebald and L.J. Fogel, editors, *Proceedings of the 3rd Annual Conference on Evolutionary Programming*, pages 296–307. World Scientific, 1994.
410. W.M. Spears. Adapting crossover in evolutionary algorithms. In J.R. McDonnell, R.G. Reynolds, and D.B. Fogel, editors, *Proceedings of the 4th Annual Conference on Evolutionary Programming*, pages 367–384. MIT Press, Cambridge, MA, 1995.
411. W.M. Spears. *Evolutionary Algorithms: the role of mutation and recombination.* Springer, 2000.
412. W.M. Spears and K.A. De Jong. An analysis of multi point crossover. In Rawlins [351], pages 301–315.
413. W.M. Spears and K.A. De Jong. On the virtues of parameterized uniform crossover. In Belew and Booker [46], pages 230–237.

414. W.M. Spears and K.A. De Jong. Dining with GAs: Operator lunch theorems. In Banzhaf and Reeves [38], pages 85–101.
415. L. Spector, E. Goodman, A. Wu, W.B. Langdon, H.-M. Voigt, M. Gen, S. Sen, M. Dorigo, S. Pezeshk, M. Garzon, and E. Burke, editors. *Proceedings of the Genetic and Evolutionary Computation Conference (GECCO-2001)*. Morgan Kaufmann, San Francisco, 2001.
416. L. Spector, W.B. Langdon, U.-M. O'Reilly, and P.J. Angeline, editors. *Advances in Genetic Programming 3*. MIT Press, Cambridge, MA, 1999.
417. N. Srinivas and K. Deb. Multiobjective optimization using nondominated sorting in genetic algorithms. *Evolutionary Computation*, 2(3):221–248, Fall 1994.
418. C.R. Stephens and H. Waelbroeck. Schemata evolution and building blocks. *Evolutionary Computation*, 7(2):109–124, 1999.
419. R. Storn and K. Price. Differential evolution - a simple and efficient adaptive scheme for global optimization over continuous spaces. Technical Report TR-95-012, ICSI, Berkeley, March 1995.
420. P.N. Suganthan, N. Hansen, J.J. Liang, K.Deb, Y.-P. Chen, A. Auger, and S. Tiwari. Problem definitions and evaluation criteria for the CEC 2005 special session on real-parameter optimization. Technical report, Nanyang Technological University, 2005.
421. P. Surry and N. Radcliffe. Innoculation to initialise evolutionary search. In T.C. Fogarty, editor, *Evolutionary Computing: Proceedings of the 1996 AISB Workshop*, pages 269–285. Springer, 1996.
422. G. Syswerda. Uniform crossover in genetic algorithms. In Schaffer [365], pages 2–9.
423. G. Syswerda. Schedule optimisation using genetic algorithms. In Davis [98], pages 332–349.
424. G. Taguchi and T. Yokoyama. *Taguchi Methods: Design of Experiments*. ASI Press, 1993.
425. R. Tanese. Parallel genetic algorithm for a hypercube. In Grefenstette [198], pages 177–183.
426. D.M. Tate and A.E. Smith. Unequal area facility layout using genetic search. *IIE transactions*, 27:465–472, 1995.
427. L. Tesfatsion. Preferential partner selection in evolutionary labor markets: A study in agent-based computational economics. In V.W. Porto, N. Saravanan, D. Waagen, and A.E. Eiben, editors, *Proc. 7th Annual Conference on Evolutionary Programming*, number 1477 in LNCS, pages 15–24. Springer, 1998.
428. S.R. Thangiah, R. Vinayagamoorty, and A.V. Gubbi. Vehicle routing and time deadlines using genetic and local algorithms. In Forrest [176], pages 506–515.
429. D. Thierens. Adaptive strategies for operator allocation. In Lobo et al. [273], pages 77–90.
430. D. Thierens and D.E. Goldberg. Mixing in genetic algorithms. In Forrest [176], pages 38–45.
431. C.K. Ting, W.M. Zeng, and T.C.Lin. Linkage discovery through data mining. *Computational Intelligence Magazine*, 5:10–13, 2010.
432. Vito Trianni. *Evolutionary Swarm Robotics – Evolving Self-Organising Behaviours in Groups of Autonomous Robots*, volume 108 of *Studies in Computational Intelligence*. Springer, 2008.
433. E.P.K. Tsang. *Foundations of Constraint Satisfaction*. Academic Press, 1993.
434. S. Tsutsui. Multi-parent recombination in genetic algorithms with search space boundary extension by mirroring. In Eiben et al. [131], pages 428–437.

435. P.D. Turney. How to shift bias: lessons from the Baldwin effect. *Evolutionary Computation*, 4(3):271–295, 1996.
436. R. Unger and J. Moult. A genetic algorithm for 3D protein folding simulations. In Forrest [176], pages 581–588.
437. J. I. van Hemert and A. E. Eiben. Mondriaan art by evolution. In Eric Postma and Marc Gyssens, editors, *Proceedings of the Eleventh Belgium/Netherlands Conference on Artificial Intelligence (BNAIC'99)*, pages 291–292, 1999.
438. C.H.M. van Kemenade. Explicit filtering of building blocks for genetic algorithms. In Voigt et al. [445], pages 494–503.
439. E. van Nimwegen, J.P. Crutchfield, and M. Mitchell. Statistical dynamics of the Royal Road genetic algorithm. *Theoretical Computer Science*, 229:41–102, 1999.
440. F.J. Varela and P. Bourgine, editors. *Toward a Practice of Autonomous Systems: Proceedings of the 1st European Conference on Artificial Life.* MIT Press, Cambridge, MA, 1992.
441. F. Vavak and T.C. Fogarty. A comparative study of steady state and generational genetic algorithms for use in nonstationary environments. In T.C. Fogarty, editor, *Evolutionary Computing*, pages 297–304. Springer, 1996.
442. F. Vavak and T.C. Fogarty. Comparison of steady state and generational genetic algorithms for use in nonstationary environments. In ICEC-96 [230], pages 192–195.
443. F. Vavak, T.C. Fogarty, and K. Jukes. A genetic algorithm with variable range of local search for tracking changing environments. In Voigt et al. [445], pages 376–385.
444. F. Vavak, K. Jukes, and T.C. Fogarty. Adaptive combustion balancing in multiple burner boiler using a genetic algorithm with variable range of local search. In Bäck [23], pages 719–726.
445. H.-M. Voigt, W. Ebeling, I. Rechenberg, and H.-P. Schwefel, editors. *Proceedings of the 4th Conference on Parallel Problem Solving from Nature*, number 1141 in Lecture Notes in Computer Science. Springer, 1996.
446. M.D. Vose. *The Simple Genetic Algorithm.* MIT Press, Cambridge, MA, 1999.
447. M.D. Vose and G.E. Liepins. Punctuated equilibria in genetic search. *Complex Systems*, 5(1):31, 1991.
448. M. Waibel, L. Keller, and D. Floreano. Genetic Team Composition and Level of Selection in the Evolution of Cooperation. *IEEE transactions on Evolutionary Computation*, 13(3):648–660, 2009.
449. J. Walker, S. Garrett, and M. Wilson. Evolving controllers for real robots: A survey of the literature. *Adaptive Behavior*, 11(3):179–203, 2003.
450. L. Wang, K.C. Tan, and C.M. Chew. *Evolutionary Robotics: from Algorithms to Implementations*, volume 28 of *World Scientific Series in Robotics and Intelligent Systems.* World Scientific, 2006.
451. P.M. White and C.C. Pettey. Double selection vs. single selection in diffusion model GAs. In Bäck [23], pages 174–180.
452. D. Whitley. Permutations. In Bäck et al. [27], chapter 33.3, pages 274–284.
453. D. Whitley, D. Goldberg, E. Cantu-Paz, L. Spector, I. Parmee, and H.-G. Beyer, editors. *Proceedings of the Genetic and Evolutionary Computation Conference (GECCO-2000).* Morgan Kaufmann, San Francisco, 2000.
454. D. Whitley, K. Mathias, S. Rana, and J. Dzubera. Building better test functions. In Eshelman [156], pages 239–246.

455. L.D. Whitley. Fundamental principles of deception in genetic search. In Rawlins [351], pages 221–241.
456. L.D. Whitley. Cellular genetic algorithms. In Forrest [176], pages 658–658.
457. L.D. Whitley, editor. *Foundations of Genetic Algorithms - 2*. Morgan Kaufmann, San Francisco, 1993.
458. L.D. Whitley, S. Gordon, and K.E. Mathias. Lamarckian evolution, the Baldwin effect, and function optimisation. In Davidor et al. [96], pages 6–15.
459. L.D. Whitley and F. Gruau. Adding learning to the cellular development of neural networks: evolution and the Baldwin effect. *Evolutionary Computation*, 1:213–233, 1993.
460. L.D. Whitley and J. Kauth. Genitor: A different genetic algorithm. In *Proceedings of the Rocky Mountain Conference on Artificial Intelligence*, pages 118–130, 1988.
461. L.D. Whitley, K.E. Mathias, and P. Fitzhorn. Delta coding: An iterative search strategy for genetic algorithms,. In Belew and Booker [46], pages 77–84.
462. L.D. Whitley and M.D. Vose, editors. *Foundations of Genetic Algorithms 3*. Morgan Kaufmann, San Francisco, 1995.
463. W. Wickramasinghe, M. van Steen, and A. E. Eiben. Peer-to-peer evolutionary algorithms with adaptive autonomous selection. In D. Thierens *et al.*, editor, *GECCO '07: Proc of the 9th conference on Genetic and Evolutionary Computation*, pages 1460–1467. ACM Press, 2007.
464. S.W. Wilson. ZCS: A zeroth level classifier system. *Evolutionary Computation*, 2(1):1–18, 1994.
465. S.W. Wilson. Classifier fitness based on accuracy. *Evolutionary Computation*, 3(2):149–175, 1995.
466. A.F.T. Winfield and J. Timmis. Evolvable robot hardware. In *Evolvable Hardware: from Practice to Applications*, Natural Computing Series, page in press. Springer, 2015.
467. D.H. Wolpert and W.G. Macready. No Free Lunch theorems for optimisation. *IEEE Transactions on Evolutionary Computation*, 1(1):67–82, 1997.
468. S. Wright. The roles of mutation, inbreeding, crossbreeding, and selection in evolution. In *Proc. of 6th Int. Congr. on Genetics*, volume 1, pages 356–366. Ithaca, NY, 1932.
469. X. Yao and Y. Liu. Fast evolutionary programming. In Fogel et al. [172].
470. X. Yao, Y. Liu, and G. Lin. Evolutionary programming made faster. *IEEE Transactions on Evolutionary Computing*, 3(2):82–102, 1999.
471. B. Yuan and M. Gallagher. Combining Meta-EAs and Racing for Difficult EA Parameter Tuning Tasks. In Lobo et al. [273], pages 121–142.
472. J. Zar. *Biostatistical Analysis.* Prentice Hall, 4th edition, 1999.
473. Zhi-hui Zhan and Jun Zhang. Adaptive particle swarm optimization. In M. Dorigo, M. Birattari, C. Blum, M. Clerc, T. Stützle, and A. Winfield, editors, *Ant Colony Optimization and Swarm Intelligence*, volume 5217 of *Lecture Notes in Computer Science*, pages 227–234. Springer, 2008.
474. Qingfu Zhang and Hui Li. MOEA/D: A Multiobjective Evolutionary Algorithm Based on Decomposition. *IEEE Transactions on Evolutionary Computation*, 11(6):712–731, 2007.
475. E. Zitzler, M. Laumanns, and L. Thiele. SPEA2: Improving the strength Pareto evolutionary algorithm for multiobjective optimization. In K.C. Giannakoglou, D.T.. Tsahalis, J. Périaux, K.D. Papailiou, and T.C. Fogarty, editors, *Evolutionary Methods for Design Optimization and Control with Applications to Industrial Problems*, pages 95–100, Athens, Greece, 2001. International Center for Numerical Methods in Engineering (Cmine).

Index